AF581716

Natural Alternatives against Bacterial Foodborne Pathogens

Natural Alternatives against Bacterial Foodborne Pathogens

Special Issue Editors

Adolfo J. Martinez-Rodriguez

Jose Manuel Silvan

MDPI • Basel • Beijing • Wuhan • Barcelona • Belgrade • Manchester • Tokyo • Cluj • Tianjin

Special Issue Editors
Adolfo J. Martinez-Rodriguez
Institute of Food Science Research
(CIAL)CSIC-UAM
Spain

Jose Manuel Silvan
Institute of Food Science Research
(CIAL)CSIC-UAM
Spain

Editorial Office
MDPI
St. Alban-Anlage 66
4052 Basel, Switzerland

This is a reprint of articles from the Special Issue published online in the open access journal *Microorganisms* (ISSN 2076-2607) (available at: https://www.mdpi.com/journal/microorganisms/special_issues/natural_alternatives_foodborne).

For citation purposes, cite each article independently as indicated on the article page online and as indicated below:

LastName, A.A.; LastName, B.B.; LastName, C.C. Article Title. *Journal Name* **Year**, *Article Number*, Page Range.

ISBN 978-3-03936-551-7 (Hbk)
ISBN 978-3-03936-552-4 (PDF)

Cover image courtesy of Jose Manuel Silván Jiménez.

Contents

About the Special Issue Editors

Adolfo J. Martinez-Rodriguez holds a 5-year degree in Biochemistry and a PhD in Chemical Sciences. He followed his PhD with a postdoctoral period at the University of Reading, UK. At present, he is Senior Scientist of the Spanish Science Research Council (CSIC), in charge of the scientific direction of the Microbiology and Biocatalysis group of the Institute of Food Science Research CIAL (UAM-CSIC). He is coauthor of 85 original scientific papers in the field of food microbiology and 10 invention patents, 5 of them licensed to food manufacturers. His current research interests are mainly focused in the areas of food microbiology and food safety in the control of human pathogens, using alternative strategies to antibiotics based on the valorization of industrial by-products as a source of bioactive extracts against fastidious pathogens such as *Campylobacter* spp. and *H. pylori*.

Jose Manuel Silvan obtained his PhD degree in Food Science and Technology at Autonoma University of Madrid (2010). Dr. Silván currently works as research scientist at the Food Microbiology and Biocatalysis Laboratory (MICROBIO) in the Institute of Food Science Research (CIAL) of the Spanish Science Research Council (CSIC). He is co-author of more than 40 peer-reviewed scientific publications in the food science and technology area. He is co-author of several book chapters and co-editor of various Special Issues of SCI indexed journals. He has participated in 21 research projects with public and private funding, and he has contributed in more than 50 national and international scientific conferences, with the recognition of two awards. Currently, his research field is focused on the finding of alternative strategies to the use of antibiotics and disinfectants for the control of pathogenic microorganisms, such as *Campylobacter* and *Helicobacter*, using food by-products and other compounds of natural origin as a source of bioactive extracts.

Editorial

Editorial for *Special Issue* "Natural Alternatives against Bacterial Foodborne Pathogens"

Adolfo J. Martinez-Rodriguez * and Jose Manuel Silvan *

Microbiology and Food Biocatalysis Group, Department of Biotechnology and Food Microbiology, Institute of Food Science Research (CIAL), CSIC-UAM-C/Nicolás Cabrera, 9. Cantoblanco Campus, Autonoma University of Madrid, 28049 Madrid, Spain

* Correspondence: adolfo.martinez@csic.es (A.J.M.-R.); jm.silvan@csic.es (J.M.S.)

Received: 7 May 2020; Accepted: 19 May 2020; Published: 20 May 2020

In recent years, increased resistance to antibiotics and disinfectants from foodborne bacterial pathogens has become a relevant consumer health issue and a growing concern for food safety authorities. In this situation, and with an apparent stagnation in the development of broad-spectrum antibiotics, research into new antibacterial agents and strategies for the control of foodborne pathogens that have good acceptability, low toxicity levels, and high sustainability is greatly demanded at present.

This *Special Issue* on "Natural Alternatives against Bacterial Foodborne Pathogens" aims to contribute to the visibility of some of these new antibacterial agents and contains eight research articles and one review, presenting different strategies potentially applicable in the control of various foodborne pathogens.

The antibacterial properties of extra virgin olive oil against different foodborne pathogens and their relationship with phenolic composition of the extract are described by Nazzaro et al. [1]. This study may contribute to the design of optimal mixtures of polyphenols with improved antibacterial efficacy. The paper by Silvan et al. [2] reports that plum extract powders gained after freeze-, vacuum- and spray-drying have promising antibacterial, antioxidant, and anti-inflammatory properties, demonstrating that the drying method selected can be an effective tool for modulating the composition, physical, and bioactive properties of plum extracts powders. The antimicrobial effect of essential oils obtained from cinnamon, marjoram, and thyme on single and dual biofilms of *Escherichia coli, Listeria monocytogenes*, *Pseudomonas putida*, and *Staphylococcus aureus* is described by Kerekes et al. [3]. These studies are the starting point for new approaches, such as encapsulation of essential oils, that could potentially reduce its organoleptic impact and increase antibacterial activity. Following a different strategy, Speranza et al. [4] proposes to exploit the in vivo metabolism of two probiotic strains (*Bifidobacterium longum* subsp. *infantis* and *Lactobacillus reuteri*) with the capacity to adhere on different surfaces (i.e., packaging materials, ceramic, plastic, paper, polymers, etc) forming a biofilm able to control the growth of pathogenic and food spoilage bacteria. This could be useful as a new biocontrol solution for different industrial applications. The probiotic functionality of a *Bacillus subtilis* strain protecting probiotic lactic acid bacteria during their exposure to unfavorable environmental conditions, such as desiccation and acid stresses, is described by Kimelman and Shemesh [5]. In addition to this protective capability, *B. subtilis* strains have demonstrated a potent antimicrobial activity against pathogenic *S. aureus*. Luis et al. [6] report the development of hydrophobic zein-based functional films incorporating licorice essential oil as new alternative materials for food packaging. These new films are biodegradable and possess antioxidant and antibacterial properties against different foodborne pathogens, making them potential alternatives to the conventional plastics used in food packaging solutions, reducing environmental pollution and increasing the shelf-life of foods. Zhang et al. [7] present the antibacterial activity of different spice extracts against several antibiotic resistant strains of foodborne pathogens. They conclude that some extracts with relevant antibacterial and antioxidant activity could have potential for use as both antibiotic alternatives in animal feeding and as a natural food preservative in the food industry. Kaewkod et al. [8] report the study of different biological properties of Kombucha tea from various kinds of tea leaves including green, oolong, and black tea. They observe that the extent of the antibacterial effect against several foodborne pathogens was related with the amount of organic acids in the beverage, indicating the great potential health benefits of Kombucha tea. Finally, Zorraquin-Peña et al. [9] present a detailed review on the main applications of silver nanoparticles as antibacterial agents for food control, as well as the current legislation concerning these materials. They also summarize the current knowledge about the impact of dietary exposure to silver nanoparticles in human health, with special emphasis on the changes that nanoparticles undergo after passing through the gastrointestinal tract and how they alter the oral and gut microbiota.

Acknowledgments: Thanks to all the authors and reviewers for their excellent contributions to this *Special Issue*. Additional thanks to the *Microorganisms* Editorial Office for their professional assistance and continuous support.

Conflicts of Interest: The editors declares no conflict of interest.

References

1. Nazzaro, F.; Fratianni, F.; Cozzolino, R.; Martignetti, A.; Malorni, L.; De Feo, V.; Cruz, A.G.; d'Acierno, A. Antibacterial Activity of Three Extra Virgin Olive Oils of the Campania Region, Southern Italy, Related to Their Polyphenol Content and Composition. *Microorganisms* **2019**, *7*, 321. [CrossRef] [PubMed]
2. Silvan, J.M.; Michalska-Ciechanowska, A.; Martinez-Rodriguez, A.J. Modulation of Antibacterial, Antioxidant, and Anti-Inflammatory Properties by Drying of *Prunus domestica* L. Plum Juice Extracts. *Microorganisms* **2020**, *8*, 119. [CrossRef] [PubMed]
3. Kerekes, E.B.; Vidács, A.; Takó, M.; Petkovits, T.; Vágvölgyi, C.; Horváth, G.; Balázs, V.L.; Krisch, J. Anti-Biofilm Effect of Selected Essential Oils and Main Components on Mono- and Polymicrobic Bacterial Cultures. *Microorganisms* **2019**, *7*, 345. [CrossRef] [PubMed]
4. Speranza, B.; Liso, A.; Russo, V.; Corbo, M.R. Evaluation of the Potential of Biofilm Formation of *Bifidobacterium longum* subsp. infantis and Lactobacillus reuteri as Competitive Biocontrol Agents against Pathogenic and Food Spoilage Bacteria. *Microorganisms* **2020**, *8*, 177. [CrossRef] [PubMed]
5. Kimelman, H.; Shemesh, M. Probiotic Bifunctionality of *Bacillus subtilis*—Rescuing Lactic Acid Bacteria from Desiccation and Antagonizing Pathogenic *Staphylococcus aureus*. *Microorganisms* **2019**, *7*, 407. [CrossRef] [PubMed]
6. Luís, Â.; Domingues, F.; Ramos, A. Production of Hydrophobic Zein-Based Films Bioinspired by The Lotus Leaf Surface: Characterization and Bioactive Properties. *Microorganisms* **2019**, *7*, 267.
7. Zhang, D.; Gan, R.-Y.; Farha, A.K.; Kim, G.; Yang, Q.-Q.; Shi, X.-M.; Shi, C.-L.; Luo, Q.-X.; Xu, X.-B.; Li, H.-B.; et al. Discovery of Antibacterial Dietary Spices That Target Antibiotic-Resistant Bacteria. *Microorganisms* **2019**, *7*, 157. [CrossRef] [PubMed]
8. Kaewkod, T.; Bovonsombut, S.; Tragoolpua, Y. Efficacy of Kombucha Obtained from Green, Oolong, and Black Teas on Inhibition of Pathogenic Bacteria, Antioxidation, and Toxicity on Colorectal Cancer Cell Line. *Microorganisms* **2019**, *7*, 700. [CrossRef] [PubMed]
9. Zorraquín-Peña, I.; Cueva, C.; Bartolomé, B.; Moreno-Arribas, M.V. Silver Nanoparticles against Foodborne Bacteria. Effects at Intestinal Level and Health Limitations. *Microorganisms* **2020**, *8*, 132.

Article

Antibacterial Activity of Three Extra Virgin Olive Oils of the Campania Region, Southern Italy, Related to Their Polyphenol Content and Composition

Filomena Nazzaro [1,*], Florinda Fratianni [1], Rosaria Cozzolino [1], Antonella Martignetti [1], Livia Malorni [1], Vincenzo De Feo [2], Adriano G. Cruz [3] and Antonio d'Acierno [1]

1 Istituto di Scienze dell'Alimentazione-Consiglio Nazionale delle Ricerche (CNR-ISA), Via Roma 64, 83100 Avellino, Italy
2 Dipartimento di Farmacia, Università di Salerno, Via Giovanni Paolo II, 132, Fisciano, 84084 Salerno, Italy
3 Instituto Federal de Educação, Ciência e Tecnologia di Rio de Janeiro (IFRJ), Departamento de Alimentos, Rio de Janeiro 20270-021, Brazil
* Correspondence: filomena.nazzaro@cnr.it; Tel.: +39-0825-299-102

Received: 22 July 2019; Accepted: 4 September 2019; Published: 5 September 2019

Abstract: Production of extra virgin olive oil (EVOO) represents an important element for the economy of Southern Italy. Therefore, EVOO is recognized as a food with noticeable biological effects. Our study aimed to evaluate the antimicrobial activity exhibited by the polyphenolic extracts of EVOOs, obtained from three varieties of *Olea europea* L. (*Ruvea antica*, *Ravece*, and *Ogliarola*) cultivated in the village of Montella, Avellino, Southern Italy. The study evaluated the inhibiting effect of the extracts against some Gram-positive and Gram-negative bacteria. Statistical analysis, used to relate values of antimicrobial activity to total polyphenols and phenolic composition, revealed a different behavior among the three EVOO polyphenol extracts. The method applied could be useful to predict the influence of singular metabolites on the antimicrobial activity.

Keywords: extra virgin olive oil; polyphenols; antimicrobial activity

---

1. Introduction

Extra virgin olive oil (EVOO) is a food extracted by the mechanical pressing of the fruits of the olive tree (*Olea europaea* L.). EVOO and other products from olive tree are central components of the Mediterranean diet, characterized, as it is well known, by a scarce intake of products of terrestrial animal origin, and, concomitantly, by a high intake of fruits, vegetables, cereals, fish, as well as by a moderate wine consumption. Fruits and vegetables, including cereals, are rich in phytochemicals, with proven protective effects in limiting several chronic diseases, such as cancer and cardiovascular illnesses. EVOO represents an important source of nutritionally and healthfully compounds, so that it is considered as a real functional food [1]. Apart from fatty acids (mainly triglycerides, fat-soluble substances and polar compounds, representing 95–98% of the whole EVOO)—pulp and seed of olive contain several other types of compounds, which are present in the final product after the extractive process. Polyphenols are probably one of the most important groups of minor polar components present in the EVOO. The biological importance of polyphenols gives rise from their numerous ascertained biochemical activities, such as the prevention of oxidation reactions to fatty acids. In addition, for this reason they contribute to the stability of the oil over time, delaying rancidity. Polyphenols are also capable of preventing and inhibiting radical-type reactions in the human body, thus limiting the formation of anomalous molecules that might alter the smooth functioning of cell membranes. Generally, EVOO is rich in polyphenols, until 1 g gallic acid equivalents (GAE)/kg of product [2]. The principal subfamilies of polyphenols detectable in the EVOO are phenolic acids, phenolic alcohols, secoridoids, lignans and flavonoids. Each of the above-mentioned subfamilies can then be differentiated from the others by chemical composition and reactivity, as well as, probably, by its organoleptic characteristics. It is therefore clear that the proportions and rate between the different polyphenols present in the EVOO considerably change its nutraceutical and sensory qualities. Olives and its derived-products, including EVOO, are capable, within certain limits, to resist against the biotic and abiotic stresses, for instance against pathogen attack, affecting the host-pathogen interaction. Such property is mainly due to the presence of polyphenols, which can also exhibit antimicrobial activity [3]. Polyphenols of EVOOs are able to inhibit in vitro, generally in a synergistic way, the growth of pathogens responsible for some intestine and respiratory diseases. Olive polyphenols could contribute in inhibiting the growth of *Helicobacter pylori* [4] and that of some foodborne pathogens, such as *Escherichia coli, Listeria monocytogenes* and *Salmonella enteriditis* [5]. EVOO demonstrated a good antimicrobial effect against *Salmonella* Typhi [6]. EVOO polyphenols are considerably absorbed (up to 95%) in humans mainly in the small intestine, where they might exert a significant local action [7]. Therein, they undergo different fate: some of them are directly absorbed; others are metabolized giving rise to other molecules, which can play a double role: act against enteropathogens, for instance, and, among other activities, improve the growth of beneficial microbes, acting as prebiotics [8,9]. Taking also into account the bioavailability of polyphenols, several authors ascertained that the use of EVOO in food might help in supporting the prevention against foodborne pathogens [5,10]. Recently, the inhibitory effect of EVOO polyphenols was demonstrated also against some oral microorganisms, such as oral streptococci, *Porphyromonas gingivalis, Fusobacterium nucleatum,* and *Parvimonas micra* [11]. In olive mill wastewater, phenolic compounds and their secoiridoid derivatives present in an ethanol fraction contribute to support the noticeable antimicrobial activity exhibited against the foodborne pathogen *Campylobacter* [12]. Cultivar, genetics, agronomic practices and climatic conditions, as well as the degree of ripening, storage conditions and fruit processing techniques are all factors that may affect the characteristics of EVOO, including the polyphenol profile and the subsequent biological properties [13,14]. The aim of our work was to evaluate the antibacterial activity exhibited by the polyphenol fraction of EVOOs, produced with the fruits of three varieties of *Olea europea* L. (*Ruvea antica, Ravece,* and *Ogliarola*) cultivated in Southern Italy. The study evaluated in particular the inhibitory effect of the extracts against several Gram-positive and Gram-negative bacterial strains. Statistical analysis correlated the antibacterial activity to the total polyphenols and to the percentage of the single components identified by a chromatographic approach within the three extracts.

2. Materials and Methods

The EVVOs used in this study were obtained by cold pressing from three varieties, *Ruvea antica, Ogliarola,* and *Ravece* of *O. europea,* grown in the village of Montella, Irpinia province, Campania region, Southern Italy. Samples of the three varieties were identified by Vincenzo De Feo, University of Salerno. Voucher specimens of the three varieties are stored in the herbarium of the University of Salerno.

2.1. Polyphenols Analysis

2.1.1. Standards and Reagents

Most of the standards used for the Ultra Pressure Liquid Chromatography (UPLC) analysis (caffeic, ferulic, *p*-coumaric, gallic, and chlorogenic acids; catechin; quercetin; 3-hydroxytyrosol, spiraeoside, oleureopin, dadzein, luteolin, naringenin, formononentin), as well as high pressure liquid chromatography (HPLC)-grade ethanol and acetonitrile were purchased from Sigma-Aldrich (Milano, Italy). Apigenin and hyperoside were purchased from Extrasynthese (Genay, France).

2.1.2. Extraction and Determination of Total Polyphenols

The extraction of polyphenols from EVOOs, necessary for the chromatographic analyses, was performed using hexane (1:1 *w*/*v*), following the method of Fratianni et al. [15]. The mixture was then charged onto cartridges SPE C_{18}, and eluted three times with methanol. The three residues were pooled, dried, re-suspended in 1 mL of methanol and filtered through a 0.20 mm filter before the analysis. Total phenolic (TP) content was determined using the Folin-Ciocalteau reagent [16]. The absorbance at λ = 760 nm was determined (Cary UV/Vis spectrophotometer, Varian, Palo Alto, CA, USA) at room temperature. A standard curve generated using gallic acid as standard was used to quantify total polyphenols.

2.1.3. Chromatographic Analysis

Polyphenol composition was obtained through ultra-high-performance liquid chromatography (UPLC) using an ACQUITY Ultra Performance system linked to a PDA 2996 photodiode array detector (Waters, Milford, MA, USA), linked to an Empower software (Waters). The analysis was performed following the method of Ombra et al. [17] at λ = 280 nm with a reversed-phase column (BEH C_{18}, 1.7 µm, 2.1 mm× 100 mm, Waters), at 30 °C, at a flow rate of 250 µL/min, and with pressure ranging from 6000 to 8000 psi. The effluent was introduced to an LC detector (scanning range 210–400 nm, resolution 1.2 nm). The injection volume was 5 µL. Phenolic compounds were identified and quantified through comparison of the peak areas on the chromatograms of samples with those of diluted standard solutions.

2.2. *Antibacterial Activity*

2.2.1. Microorganisms and Culture Conditions

Five Gram-positive (*Bacillus cereus* DSM 4313, *Bacillus cereus* DSM 4384, *Staphylococcus aureus* DMS 25923, *Enterococcus faecalis* DSM 2352 and *Listeria innocua* DSM 20649) and two Gram-negative (*Escherichia coli* DSM 8579, and *Pseudomonas aeruginosa* ATCC 50071) bacterial strains were cultured for 18 h in Luria Bertani (LB) broth (Sigma, Milano, Italy) at 37 °C and 80 rpm (Corning LSE, Pisa, Italy).

2.2.2. Determination of the Antibacterial Susceptibility by Agar Diffusion

The agar diffusion test was performed following the method of Fratianni et al. [18] with some modifications. Microbial suspensions (1×10^7 colony-forming units (cfu)/mL) were spread on LB agar plates in sterile conditions. Different amounts of extracts (2.5 and 4.9 µg) were spotted on the inoculated plates. After 10 min in sterile conditions, plates were incubated at 37 °C for 24 h. The diameter of the clear zone shown on plates (inhibition zone) was accurately measured ("Extra steel Caliper mod 0289", mm/inch reading scale, precision 0.05 mm, Mario De Maio, Milan, Italy). Sterile dimethylsulfoxide (DMSO, Sigma Aldrich Italy, Milano, Italy) and tetracycline (7 µg; Sigma Aldrich Italy) served as the negative and positive control, respectively. The experiments were performed in triplicate and averaged.

2.2.3. Minimal Inhibitory Concentration (MIC)

The resazurin microtiter-plate assay [19] was used to evaluate the MIC. Samples were dissolved in sterile DMSO; then, they were distributed in a multiwell plate with different volumes of sterile Muller-Hinton broth (Sigma Aldrich Italy) previously prepared. Two-fold serial dilutions were performed to have 50 µL of the test material in serially descending concentrations in each well. A 35 µL amount of 3.3 × strength iso-sensitized broth and 5 µL of resazurin, used as indicator solution, were added to achieve a final volume/well of 240 µL. Finally, 10 µL of bacterial suspension was added to each well to reach a concentration of about 5×10^5 cfu/mL. Sterile DMSO and ciprofloxacin (Sigma Aldrich Italy, prepared dissolving 1 mg/mL in DMSO) were used as negative and positive control, respectively. Multiwell plates were prepared in triplicate and incubated at 37 °C for 24 h. The lowest concentration at which a color change occurred (from dark purple to colorless) revealed the MIC value.

2.3. Statistical Analysis

Data were expressed as mean ± standard deviation of triplicate measurements. The PC software "Excel Statistics" was used for the calculations. The analysis correlated the values of antibacterial activity, specifically to the inhibition zone data, to total polyphenols and phenolic composition, using the free software environment for statistical computing and graphics R (https://www.r-project.org/) [15].

3. Results and discussion

3.1. Antibacterial Activity of the Extracts

The antibacterial capability of the polyphenol (PF) extracts of *Ogliarola*, *Ravece*, and *Ruvea antica* EVOOs was assayed against different Gram-positive and Gram-negative bacteria, through the inhibition zone test and the determination of the Minimal Inhibitory Concentration (MIC). Results are shown in Tables 1 and 2 respectively.

Table 1. Antibacterial activity evaluated through the inhibition zone test of the three polyphenol (PF) extracts of *Ogliariola*, *Ravece* and *Ruvea antica* EVOOs, against different pathogens. The test was performed using 2.5 and 4.9 μg of extract. Data are expressed in mm. Results are shown as mean (± SD) (n = 3). For details, see Materials and Methods.

	'Ogliarola'		'Ravece'		'Ruvea Antica'		Tetracycline
	2.5 μg	4.9 μg	2.5 μg	4.9 μg	2.5 μg	4.9 μg	7 μg
E. coli	7.30 (±0.57)	13.30 (±0.57)	7.00 (±0.57)	13.67 (±0.28)	5.30 (±0.52)	10.00 (±0.00)	12.67 (±0.57)
L. innocua	5.67 (±0.57)	10.67 (±0.57)	6.67 (0.57)	13.33 (±0.57)	4.30 (±0.57)	9.30 (±0.57)	10.33 (±0.50)
S. aureus	7.30 (±0.57)	11.67 (±0.57)	0.00 (±0.00)	0.00 (±0.00)	6.67 (±0.57)	12.67 (±0.57)	6.67 (±0.57)
B. cereus 4313	10.67 (±1.14)	18.33 (±0.57)	9.67 (±0.57)	17.33 (±1.15)	6.33 (±0.57)	11.67 (±0.57)	9.67 (±0.57)
B. cereus 4384	7.67 (±0.57)	13.67 (±0.57)	7.67 (±0.57)	17.30 (±1.14)	0.00 (±0.00)	0.00 (±0.00)	8.30 (±1.05)
P. aeruginosa	6.33 (±0.57)	11.33 (±0.57)	8.67 (±0.57)	16.33 (±0.57)	4.33 (0.57)	6.67 (±0.57)	10.00 (±0.00)
E. faecalis	5.67 (±0.57)	11.33 (±0.57)	7.67 (±0.57)	17.33 (±1.14)	0 00 (±0.00)	0.00 (±0.00)	12.33 (±0.57)

Table 2. Minimal Inhibitory Concentration (MIC, μg/mL) of the PF extracts of 'Ogliarola', 'Ravece' and 'Ruvea antica' EVOOs, evaluated through the resazurin test, as reported in the Materials and Methods section.

	Ogliarola	*Ravece*	*Ruvea Antica*
B. cereus 4313	1.00	1.00	1.00
B. cereus 4384	1.00	1.00	2.00
E.coli	1.00	1.00	2.00
P. aeruginosa	1.00	1.00	2.00
S. aureus	1.00	>15.00	2.00
L. innocua	2.00	2.00	2.00
E. faecalis	2.00	2.00	>10.00

The minimum concentration necessary to inhibit the growth of the pathogenic tester strains was low for all the PF extracts, usually equal to 1–2 μg, except when PF of *Ravece* were tested against *S. aureus* (MIC > 15 μg), and when those of *Ruvea antica* were assayed against *E. faecalis* (MIC > 10 μg). This confirms that polyphenols present in the EVOO have a general capacity to inhibit the growth of pathogenic or unwanted microorganisms [3]. Therefore, different in vitro studies demonstrated that some polyphenols from olive oil are able to inhibit the growth of different bacteria, including those responsible for some respiratory infection and intestinal diseases, as well as against bacteria, such as *Helicobacter pylori*, one of the agents of peptic ulcers and some types of cancer [4,20].

In general, 4.9 μg of the PF extract from *Ogliarola* were very effective in inhibiting the microbial growth of all the strains considered, with inhibition zone not lesser than 10.67 (against *L. innocua*) up to 18.33 mm (against *B. cereus* 4313). Overall, 4.9 μg of the polyphenol extract from *Ravece* produced inhibition zones also superior to 17 mm (17.33 mm, against *B. cereus* 4313 and *E. faecalis*). 4.9 μg of PF extract from *Ruvea antica* resulted less effective, producing zones not greater than 12.67 mm. All three EVOO PF extracts were effective in inhibiting the growth of *E. coli*, producing (with 4.9 μg of PF extracts from *Ravece* and *Ogliarola*) inhibition zones up to 13 mm. This result, in our opinion, could find an interesting practical application. *E. coli* is the most frequent cause of urinary tract infections. Like other *E. coli* pathotypes, the strain used in our experiments differs from the commensal *E. coli*, due to the presence of some virulence factors, which can concur, with other microbial systems, to increase its resistance against conventional antibiotics, to form biofilm, as well as to contaminate food or medical support (e.g., catheters), with difficulty to eradicate the infection and serious damage to health [21]. Thus, the capability of EVOO polyphenols to avoid the growth of this pathogen strain could be exploited not only for the EVOO per se, or for the great bioavailability of EVOO PFs, but also taking into account that the EVOO by-products are rich in polyphenols, which can convert them from a problem for the environment to a resource of biomolecules of high added value, potentially useful for food and pharmaceutical purposes. Therefore, other olive by-products, such as leaves demonstrated activity against different species of pathogens, including those used in our experiments [22]. The three PF extracts were also capable of inhibiting the growth of *Ps. aeruginosa*. Such microorganism, similar to *E. coli*, not only is a well-known pathogen, but it is also capable to form biofilm, increasing its resistance to the conventional drugs [23]. The effect was well visible, so that we measured inhibition halos until 8.67 mm just using 2.5 μg. In both cases, the extracts *Ogliarola* and *Ravece* were more effective than those of *Ruvea antica* in inhibiting the growth of the strain; in particular, 2.5 μg of PF extract of *Ravece* were twice as effective as that of *Ruvea antica* against *Ps. aeruginosa*; 4.9 μg of *Ravece* PF extracts were even three times more effective than the *Ruvea antica* ones. The different effectiveness exhibited by the extracts against the two strains of *B. cereus* (DSM 4313 and DSM 4384) proved once again that the resistance/sensitivity of a microorganism to a natural extract or to a singular compound might be not only linked to the genera or species but, in some cases, it might even be strain-specific [24,25].

3.2. Statistical Analysis

Some of the individual phenolic compounds present in the EVOOs extracts were identified and quantified by UPLC. However, the choice to evaluate the antibacterial activity of the entire extracts was taken for different reasons. First, the antibacterial activity of phenolic compounds is generally well-known [26–31]. Moreover, PF extracts might exhibit more beneficial effects than their individual constituents, which can change own properties in the presence of other compounds present in the extracts [32]. As said by Liu [33], the health benefits of fruits and vegetables give rise from synergistic effects of phytochemicals and the advantages on human health of a diet rich in fruits and vegetables is attributed to the complex mixture of phytochemicals present in whole foods. This explains why generally no individual antibacterial effect can substitute the combination of natural phytochemicals to achieve the health benefits [34]. Thus, we statistically correlated the total polyphenols and individual molecules to the antibacterial activity exhibited by the EVOO extracts. The correlation between total polyphenols and the average antibacterial activity resulted high (=0.85). We identified 10 polyphenols through UPLC analysis, based on the retention time of corresponding standards. For all of them, we calculated the percentage present in each extract. Data on polyphenol composition are reported in Table 3. The statistical approach allowed us to divide such molecules into different groups, with respect to their potential influence on the average antibacterial activity of the extracts. Correlation coefficients (Corr-coeffs) are reported in Table 4. In the first group, we found that flavonol quercetin and isoflavone formononetin, which Corr-coeffs (0.94 and 0.97, respectively) seemed to let us foresee by the whole their highest influence on the antibacterial activity with respect to the other molecules. Other two polyphenols, flavanone naringenin and the secoiridoid oleuropein exhibited lower Corr-coeffs (0.55 and 0.47, respectively).

Taking into account the percentage of the two molecules in the extracts, it is possible to hypothesize for this other group a little bit of predominance of correlation between oleuropein and the average antibacterial activity of the 'Ravece' extract (Figure 1, left) and between naringenin on the average antibacterial activity exerted by the 'Ogliarola' extract (Figure 1, right).

Table 3. Polyphenol composition, obtained by Ultra Pressure Liquid Chromatography (UPLC), of the three PF extracts of *Ogliarola, Ravece* and *Ruvea antica* EVOOs. The data are reported as percentage of total polyphenols.

Polyphenols (%)	'Ogliarola'	'Ravece'	'Ruvea Antica'
Gallic acid	0.00	0.00	0.00
3 Hydroxytirosol	1.86	0.43	1.10
Catechin	1.08	0.00	0.43
p-Coumaric acid	0.00	0.28	0.11
Quercetin-4-glucoside (spiraeoside)	9.48	0.00	5.75
Oleuropein	15.77	5.92	12.82
Dadzein	4.13	0.00	2.36
Luteolin	0.00	6.22	1.57
Quercetin	24.06	18.03	10.61
Apigenin	0.00	0.00	3.18
Naringenin	3.99	6.57	6.49
Formononentin	4.45	4.81	2.27

Table 4. Correlation coefficients between the potential average antibacterial activity and polyphenols identified in the extracts of *Ogliarola, Ravece* and *Ruvea antica* EVOOs. The analysis was elaborated with respect to the percentage of each molecule present in the extracts and in an independent way with respect to the pathogens.

Polyphenols	Corr-Values
Formononentin	0.97
Quercetin	0.94
Naringenin	0.55
Oleuropein	0.47
Luteolin	0.37
Catechin	0.35
p-Coumaric acid	0.33
Dadzein	0.28
Spiraeoside	0.27
Apigenin	−0.34

The correlation between another group of polyphenols and the antibacterial activity of the extracts was still less strict; thus, flavone luteolin (Corr-coeff = 0.37) and the hydroxycinnamic *p*-coumaric acid (Corr-coeff = 0.33) seemed to break the antibacterial activity of the extract *Ogliarola*. Concurrently, isoflavone dadzein (Corr-coeff = 0.28) and flavonol spiraeoside (Corr-coeff = 0.27) did not seem to enhance that of the extract *Ravece*. The other flavone apigenin exhibited a negative coefficient of correlation (Corr-coeff = −0.34). This metabolite is a known antibacterial compound [34,35]. However, in some cases its effect could be nil against some microorganisms [36].

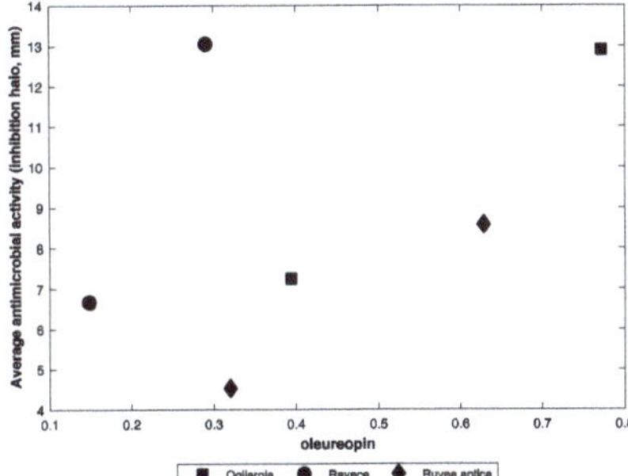

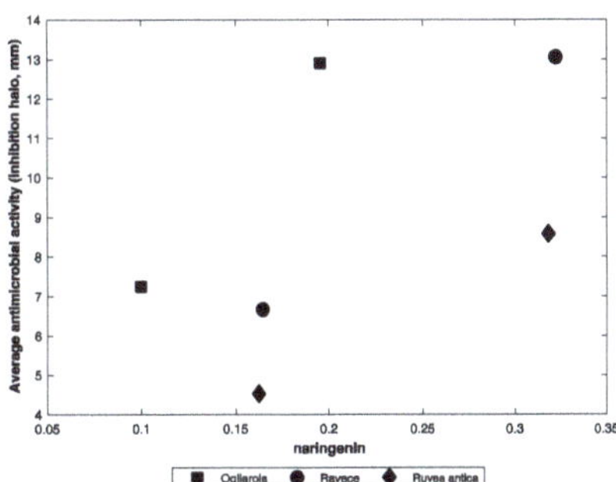

Figure 1. Average antibacterial activity exerted by the three PF extracts vs. oleuropein (**left**) and vs. naringenin (**right**). On X it is reported the amount (in μg) of the molecules present in 2.5 and 4.9 μg of the PF extracts tested.

The statistical approach was also applied to evaluate the correlation between the singular molecules and the antibacterial activity with respect to the microorganisms. Table 5 reports the coefficients of correlation.

Table 5. Correlation coefficients between the potential antibacterial activity and polyphenols identified in the extracts of 'Ogliarola', 'Ravece' and 'Ruvea antica' EVOOs, with respect to different pathogens. The analysis was elaborated with respect to the percentage of each molecule present in the extracts, taking into account the amounts (2.5 μg and 4.9 μg) of the extracts used to determine the antibacterial activity of the extracts against different pathogens. BC: *Bacillus cereus* (strains DSM 4313 and DSM 4384); EC: *Escherichia coli*; LI: EF: *Enterococcus faecalis*; *Listeria innocua*; SA: *Staphylococcus aureus*; PA: *Pseudomonas aeruginosa*.

	Microorganisms						
Polyphenol	**BC 4313**	**BC 4384**	**EC**	**EF**	**LI**	**SA**	**PA**
Formononentin	0.97	0.95	0.94	0.91	0.91	−0.16	0.95
Quercetin	0.96	0.93	0.90	0.75	0.77	0.18	0.74
Naringenin	0.47	0.57	0.65	0.26	0.78	0.02	0.55
Oleuropein	0.50	0.53	0.51	−0.09	0.33	0.89	0.00
Luteolin	0.30	0.33	0.39	0.62	0.59	−0.76	0.73
Catechin	0.41	0.38	0.33	−0.04	0.086	0.80	−0.10
p-Coumaric acid	0.25	0.30	0.36	0.52	0.58	−0.69	0.66
Dadzein	0.34	0.33	0.29	−0.19	0.06	0.90	−0.19
Spiraeoside	0.32	0.32	0.27	−0.21	0.05	0.91	−0.21
Apigenin	−0.38	−0.27	−0.21	−0.75	−0.15	0.56	−0.51
3-Hydroxytyrosol	0.51	0.51	0.47	−0.01	0.25	0.84	0.00

With respect to the strains used in the agar diffusion test, we could suppose a noticeable inhibitory effect of formononentin and quercetin against *B. cereus*. In fact, both strains of *B. cereus* (DSM 4313 and DSM 4384) seemed to be strongly inhibited by the presence of these two metabolites (Corr-coeffs = 0.97 and 0.95, respectively); concurrently, quercetin seemed to prevent the bacterial growth too (Corr-coeffs = 0.96 and 0.93, respectively). A similar effect was hypothesized against *E. coli* (Corr-coeffs = 0.94 and 0.90, respectively) and against *E. faecalis* (Corr-coeffs = 0.91 and 0.75, respectively). Thus, for instance, if formononentin seemed to confirm its influence also against *Ps. aeruginosa* (Corr-coeff = 0.95) and *L. innocua* (Corr-coeff = 0.91), on the other hand the effect of quercetin versus these two microorganisms seemed to be less effective (Corr-coeffs = 0.74 and 0.77, respectively). Therefore, other studies demonstrated a limited inhibitory effect of quercetin against *Ps. aeruginosa* [37]. A potential inhibitory effect exhibited also by luteolin (Corr-coeff = 0.73) and *p*-coumaric acid (Corr-coeff = 0.66) against *Ps. aeruginosa* was observed indeed. At the same time naringenin (Corr-coeff = 0.78), luteolin (Corr-coeff = 0.59) and *p*-coumaric acid (Corr-coeff = 0.58) would concur in influencing, although with minor efficacy, the potential antibacterial activity of the extracts against *L. innocua*. The potential behavior exhibited by metabolites on the antibacterial activity-hypothesized through such approach- seemed to be completely different when we considered *S. aureus*. In fact, by the analysis of correlation coefficients we could hypothesize that other metabolites in place of formononentin and quercetin may have contributed to the antibacterial activity of the extracts, in particular spiraeoside, dadzein, and catechin (Corr-coeffs = 0.91; 0.90 and 0.80, respectively). Moreover, this was the unique case in which oleuropein (one of the most important and known metabolites characterizing the EVOO polyphenols) seemed to have contributed to the antibacterial activity of the extracts (Corr-coeff = 0.89). Therefore, oleuropein as well as 3-hydroxytirosol (which in our case showed a correlation coefficient of 0.84) have antibacterial activity against *S. aureus*, as demonstrated by Bisignano et al. [38]. Concurrently, statistics confirmed the controversial behavior exhibited by 3-hydroxytirosol that was active against *S. aureus* but had lower effect (Corr-coeff = 0.47) against *E. coli*, corroborating the indications given by other studies [39]. The fact that the *Ravece* extract did not contain dadzein might suggest that such metabolite in particular affected the resistance of *S. aureus*. In fact, as shown in Table 2, the MIC *Ravece* extract *versus S. aureus* was higher than 10 μg and much lower in the case of the other two extracts. The absence of catechin, which gave a correlation coefficient of 0.80 and the concurrent presence of luteolin (6.22% in *Ravece*, Corr-coeff = −0.76) could have contributed to its higher MIC value. Concomitantly, the presence of apigenin found only in the *Ruvea antica* extract with the most negative coefficient of correlation (= −0.75) would seem to support its influence on the resistance of *E. faecalis* versus that extract, as indicated by the MIC value and by the results of the inhibition zone test.

4. Conclusions

The polyphenol fraction present in EVOO oil confirms once again its antibacterial properties. The different qualitative and quantitative profile of polyphenols present in a PF extract can affect in a different way its antibacterial effectiveness. The statistical method herein applied is easy and useful to predict the synergistic effect of polyphenols and the influence that each of them has—based also on their amount—on the activity of the whole extract. In a future perspective this could be a basis of possible complementary studies, for example, to formulate ideal drugs of natural origin, composed of optimal mixtures of polyphenols which are able to exercise with the minimum effort (in terms of quantity) and the maximum result (against the greatest number of pathogens) their antibacterial efficacy.

Author Contributions: Conceptualization and Investigation, F.N., F.F., A.G.C. and A.d.A.; methodology, F.N., F.F., R.C., L.M., A.M., V.D.F.; Formal analysis, A.d.A.; writing—original draft preparation, F.N., A.d.A., V.D.F.; writing—review and editing, F.N., A.d.A., V.D.F.; funding acquisition, F.N.

Funding: The Campania Regional Council, Italy within the Project funded this research: "Salvaguardia della Biodiversità Vegetale della Campania" ("SALVE"), PSR 2007–2013, mis 214, action f2.

Conflicts of Interest: The authors declare no conflict of interest.

References

1. Ray, N.B.; Hilsabeck, K.D.; Karagiannis, T.C.; McCord, D.E. Bioactive Olive Oil Polyphenols in the Promotion of Health. In *The Role of Functional Food Security in Global Health*; Singh, R.B., Watson, R.R., Takahashi, T., Eds.; Elsevier: Amsterdam, The Netherlands, 2019; pp. 623–637.
2. Gorzynik-Debicka, M.; Przychodzen, P.; Cappello, F.; Kuban-Jankowska, A.; Marino Gammazza, A.; Knap, N.; Wozniak, M.; Gorska-Ponikowska, M. Potential health benefits of olive oil and plant polyphenols. *Int. J. Mol. Sci.* **2018**, *19*, 686. [CrossRef] [PubMed]
3. Capasso, R.; Evidente, A.; Schivo, L.; Orru, G.; Marcialis, M.A.; Cristinzio, G. Antibacterial polyphenols from olive oil mill waste waters. *J. Appl. Bacteriol.* **1995**, *79*, 393–398. [CrossRef] [PubMed]
4. Romero, C.; Medina, E.; Vargas, J.; Brenes, M.; De Castro, A. In vitro activity of olive oil polyphenols against *Helicobacter pylori*. *J. Agric Food Chem.* **2007**, *55*, 680–686. [CrossRef] [PubMed]

5. Karaosmanoglu, H.; Soyer, F.; Ozen, B.; Tokatli, F. Antimicrobial and antioxidant activities of Turkish extra virgin olive oils. *J. Agric. Food Chem.* **2010**, *58*, 8238–8245. [CrossRef] [PubMed]
6. Gabriel, P.O.; Aribisala, J.O.; Oladunmoye, M.K.; Arogunjo, A.O.; Ajayi-Moses, O.B. Therapeutic effect of goya extra virgin olive oil in albino rat oro-gastrically dosed with *Salmonella* Typhi. *South Asian J. Res. Microbiol.* **2019**, *3*, 1–9.
7. Rubio, L.; Macia, A.; Castell-Auvi, A.; Pinent, M.; Blay, M.T.; Ardevol, A.; Romero, M.P.; Motilva, M.J. Effect of the co-occurring olive oil and thyme extracts on the phenolic bioaccessibility and bioavailability assessed by in vitro digestion and cell models. *Food Chem.* **2014**, *149*, 277–284. [CrossRef] [PubMed]
8. Deiana, M.; Serra, G.; Corona, G. Modulation of intestinal epithelium homeostasis by extra virgin olive oil phenolic compounds. *Food Funct.* **2018**, *9*, 4085–4099. [CrossRef] [PubMed]
9. Nazzaro, F.; Fratianni, F.; d'Acierno, A.; Coppola, R. Gut Microbiota and Polyphenols: A Strict Connection Enhancing Human Health. In *Advances in Food Biotechnology*; Ravishankar Rai, V., Ed.; John Wiley & Sons Ltd.: Chichester, UK, 2015; pp. 335–350.
10. Cicerale, S.; Lucas, L.J.; Keast, R.S.J. Antimicrobial, antioxidant and anti-inflammatory phenolic activities in extra virgin olive oil. *Curr. Opin. Biotechn.* **2012**, *23*, 129–135. [CrossRef]
11. Karygianni, L.; Cecere, M.; Argyropoulou, A.; Hellwig, E.; Skaltsounis, A.L.; Wittmer, A.; Tchorz, J.P.; Al-Ahmad, A. Compounds from *Olea europaea* and *Pistacia lentiscus* inhibit oral microbial growth. *BMC Compl. Altern. Med.* **2019**, *19*, 51. [CrossRef]
12. Manuel Silvana, J.; Pinto-Bustillos, M.A.; Vásquez-Ponce, P.; Prodanov, M.; Martinez-Rodriguez, A.J. Olive mill wastewater as a potential source of antibacterial and anti-inflammatory compounds against the food-borne pathogen *Campylobacter*. *Inn. Food Sci. Em. Technol.* **2019**, *51*, 177–185. [CrossRef]
13. Lazzez, A.; Perri, E.; Caravita, M.A.; Khlif, M.; Cossentini, M. Influence of olive maturity stage and geographical origin on some minor components in virgin olive oil of the Chemlali variety. *J. Agric. Food Chem.* **2008**, *56*, 982–988. [CrossRef] [PubMed]
14. Rotondi, A.; Bendini, A.; Cerretani, L.; Mari, M.; Lercker, G.; Toschi, T.G. Effect of olive ripening degree on the oxidative stability and organoleptic properties of cv. Nostrana di Brisighella extra virgin olive oil. *J. Agric. Food Chem.* **2004**, *52*, 3649–3654. [CrossRef] [PubMed]
15. Fratianni, F.; Cozzolino, R.; Martignetti, A.; Malorni, L.; d'Acierno, A.; De Feo, V.; Cruz, A.G.; Nazzaro, F. Biochemical composition and antioxidant activity of three extra virgin olive oils from the Irpinia province, Southern Italy. *Food Sci. Nutr.* **2019**, in press. [CrossRef]
16. Singleton, V.L.; Rossi, J.A. Colorimetry of total phenolics with phosphomolybdic-phosphotungstic acid reagents. *Am. J. Enol. Vitic.* **1965**, *16*, 144–158.
17. Ombra, M.; d'Acierno, A.; Nazzaro, F.; Riccardi, R.; Spigno, P.; Zaccardelli, M.; Pane, C.; Maione, M.; Fratianni, F. Phenolic composition and antioxidant and antiproliferative activities of the extracts of twelve common bean (*Phaseolus vulgaris* L.) endemic ecotypes of Southern Italy before and after cooking. *Oxid. Med. Cell Longev.* **2016**, *2016*, 1398298. [CrossRef] [PubMed]
18. Fratianni, F.; Ombra, M.N.; Cozzolino, A.; Riccardi, R.; Spigno, P.; Tremonte, P.; Coppola, R.; Nazzaro, F. Phenolic constituents, antioxidant, antimicrobial and anti-proliferative activities of different endemic Italian varieties of garlic (*Allium sativum* L.). *J. Funct. Foods* **2016**, *21*, 240–248. [CrossRef]
19. Sarker, S.D.; Nahar, L.; Kumarasamy, Y. Microtitre plate-based antibacterial assay incorporating resazurin as an indicator of cell growth, and its application in the in vitro antibacterial screening of phytochemicals. *Methods* **2007**, *42*, 321–324. [CrossRef] [PubMed]
20. Medina, E.; de Castro, A.; Romero, C.; Brenes, M. Comparison of the concentrations of phenolic compounds in olive oils and other plant oils: Correlation with antimicrobial activity. *J. Agric. Food Chem.* **2006**, *54*, 4954–4961. [CrossRef]
21. Nazzaro, F.; Fratianni, F.; d'Acierno, A.; De Feo, V.; Ayala Zavala, F.J.; Cruz, A.G.; Granato, D.; Coppola, R. Effect of Polyphenols on Microbial Cell-Cell Communications. In *Quorum Sensing*; Tommonaro, G., Ed.; Academic Press: New York, NY, USA, 2019; pp. 195–223.
22. Sudjana, A.N.; D'Orazio, C.; Ryan, V.; Rasool, N.; Ng, J.; Islam, N.; Riley, T.V.; Hammer, K.A. Antimicrobial activity of commercial *Olea europaea* (olive) leaf extract. *Int. J. Antimicrob. Agents* **2009**, *33*, 461–463. [CrossRef]
23. Nazzaro, F.; Fratianni, F.; Coppola, R. Quorum sensing and phytochemicals. *Int. J. Mol. Sci.* **2013**, *14*, 12607–12619. [CrossRef]

24. Cerulli, A.; Lauro, G.; Masullo, M.; Cantone, V.; Olas, B.; Kontek, B.; Nazzaro, F.; Bifulco, G.; Piacente, S. Cyclic diarylheptanoids from *Corylus avellana* green leafy covers: Determination of their absolute configurations and evaluation of their antioxidant and antimicrobial activities. *J. Nat. Prod.* **2017**, *80*, 1703–1713. [CrossRef] [PubMed]
25. Ruparelia, J.P.; Chatterjee, A.K.; Duttagupta, S.P.; Mukherji, S. Strain specificity in antimicrobial activity of silver and copper nanoparticles. *Acta Biomat.* **2008**, *4*, 707–718. [CrossRef] [PubMed]
26. Pereira, J.A.; Pereira, A.P.G.; Ferreira, I.C.F.R.; Valentão, P.; Andrade, P.B.; Seabra, R.; Estevinho, L.; Bento, A. Table olives from Portugal: Phenolic compounds, antioxidant potential and antimicrobial activity. *J. Agric. Food Chem.* **2006**, *54*, 8425–8431. [CrossRef] [PubMed]
27. Proestos, C.; Chorianopoulos, N.; Nychas, G.J.E.; Komaitis, M. RP-HPLC analysis of the phenolic compounds of plant extracts. Investigation of their antioxidant capacity and antimicrobial activity. *J. Agric. Food Chem.* **2005**, *53*, 1190–1195. [CrossRef] [PubMed]
28. Rauha, J.P.; Remes, S.; Heinonen, M.; Hopia, A.; Kähkönen, M.; Kujala, T.; Pihlaja, K.; Vuorela, H.; Vuorela, P. Antimicrobial effects of Finnish plant extracts containing flavonoids and other phenolic compounds. *Int. J. Food Microbiol.* **2000**, *56*, 3–12. [CrossRef]
29. Zhu, X.; Zhang, H.; Lo, R. Phenolic compounds from the leaf extract of artichoke (*Cynara scolymus* L.) and their antimicrobial activities. *J. Agric. Food Chem.* **2004**, *52*, 7272–7278. [CrossRef] [PubMed]
30. Puupponen-Pimia, R.; Nohynek, L.; Meier, C.; Kähkönen, M.; Heinonen, M.; Hopia, A.; Oksman-Caldentey, K.-M. Antimicrobial properties of phenolic compounds from berries. *J. Appl. Microbiol.* **2001**, *90*, 494–507. [CrossRef]
31. Pereira, A.P.; Ferreira, I.C.F.R.; Marcelino, F.; Valentão, P.; Andrade, P.B.; Seabra, R.; Estevinho, L.; Bento, A.; Pereira, J.A. Phenolic compounds and antimicrobial activity of olive (*Olea europaea* L. Cv. Cobrançosa) leaves. *Molecules* **2007**, *12*, 1153–1162. [CrossRef]
32. Borchers, A.T.; Keen, C.L.; Gerstiwin, M.E. Mushrooms, tumors, and immunity: An update. *Exp. Biol. Med.* **2004**, *229*, 393–406. [CrossRef]
33. Liu, R.H. Health benefits of fruits and vegetables are from additive and synergistic combination of phytochemicals. *Am. J. Clin. Nutr.* **2003**, *78*, 517S–520S. [CrossRef]
34. Cushnie, T.; Lamb, A.J. Antimicrobial activity of flavonoids. *Int. J. Antimicrob. Agents* **2005**, *26*, 343–356. [CrossRef] [PubMed]
35. Khanna, P.; Sharma, O.P.; Sehgal, M.; Bhargava, C.; Jain, M.; Goswami, A.; Singhvi, S.; Gupta, U.; Agarwal, R.; Sharma, P.; et al. Antimicrobial principles from tissue culture of some plant species. *Indian J. Pharm. Sci.* **1980**, *4*, 113–117.
36. Basile, A.; Sorbo, S.; Giordano, S.; Ricciardi, L.; Ferrara, S.; Montesano, D.; Castaldo Cobianchi, R.; Vuotto, M.L.; Ferrara, L. Antibacterial and allelopathic activity of extract from *Castanea sativa* leaves. *Fitoterapia* **2000**, *71*, S110–S116. [CrossRef]
37. Sakharkar, M.K.; Jayaraman, P.; Soe, W.M.; Chow, V.T.K.; Sing, L.C.; Sakharkar, K.R. In vitro combinations of antibiotics and phytochemicals against *Pseudomonas aeruginosa*. *J. Microbiol. Immunol. Infect.* **2009**, *42*, 364–370. [PubMed]
38. Bisignano, G.; Tomaino, A.; Lo Cascio, R.; Crisafi, G.; Uccella, N.; Sajia, A. On the *in-vitro* antimicrobial activity of oleuropein and hydroxytyrosol. *J. Pharm. Pharmacol.* **1999**, *51*, 971–974. [CrossRef] [PubMed]
39. Medina-Martínez, M.S.; Truchado, P.; Castro-Ibáñez, I.; Allende, A. Antimicrobial activity of hydroxytyrosol: A current controversy. *Biosci. Biotechn. Biochem.* **2016**, *80*, 801–810. [CrossRef] [PubMed]

Article

Modulation of Antibacterial, Antioxidant, and Anti-Inflammatory Properties by Drying of *Prunus domestica* L. Plum Juice Extracts

Jose Manuel Silvan [1,*], Anna Michalska-Ciechanowska [2] and Adolfo J. Martinez-Rodriguez [1,*]

1 Microbiology and Food Biocatalysis Group, Department of Biotechnology and Food Microbiology, Institute of Food Science Research (CIAL), CSIC-UAM-C/Nicolás Cabrera, 9. Cantoblanco Campus, Autonoma University of Madrid, 28049 Madrid, Spain

2 Department of Fruit, Vegetable and Plant Nutraceutical Technology, the Faculty of Biotechnology and Food Science, Wrocław University of Environmental and Life Sciences, Chełmońskiego 37, 51-630 Wrocław, Poland; anna.michalska@upwr.edu.pl

* Correspondence: jm.silvan@csic.es (J.M.S.); adolfo.martinez@csic.es (A.J.M.-R.); Tel.: +34-91-001-7900 (J.M.S.); +34-91-001-7964 (A.J.M.-R.)

Received: 16 December 2019; Accepted: 13 January 2020; Published: 15 January 2020

Abstract: The consumption of plums in a fresh form is seasonal, therefore the transformation of plum juice extracts into powdered form is a good alternative for its longer availability throughout the year. The drying process can moderate the physical and chemical properties of the plum extracts, thus, this study examined the changes in biological activity, i.e., antibacterial, antioxidant, and anti-inflammatory properties moderated by freeze, vacuum, and spray drying. It was suggested that the drying processes and the applied parameters might moderate the content of polyphenolic compounds in the powders, which influence the different levels of growth inhibition against the foodborne pathogens (17% to 58% of inhibition), demonstrating a strain-dependent effect. These powders could also induce cellular protection against oxidative stress by preventing intracellular reactive oxygen species (ROS) accumulation (23% to 37% of reduction), but the level of antioxidant capacity may be determined by the conditions applied during the drying process. Moreover, plum extract powders exhibited a greater anti-inflammatory capacity (24% to 39% of inhibition), which would be influenced both, by the type of treatment used and by the temperature used in each treatment. The results demonstrate that the selection of the drying method can be an effective tool for modulating the composition, physical, and bioactive properties of plum extracts powders.

Keywords: *Prunus domestica* L.; plum extracts; drying; polyphenolics; bioactive properties; antibacterial; antioxidant anti-inflammatory

---

1. Introduction

Diets rich in fruits are beneficial to human health because of their polyphenolic compound content. In this regard, plums (*Prunus domestica* L.) represent an excellent source of such components, which can contribute significantly to the prevention of several diseases [1]. This fruit is cultivated all over the world and its production in the last 10 years has exceeded 11 million tons [2]. Plum is a seasonal fruit, and the harvest period and the period of supply of fresh fruit are relatively short. Therefore, plums cannot be consumed fresh throughout the year, thus, the development of new dried powder products, obtained by drying industrial techniques, offers an alternative for the consumption at any season of the year. Plum powders can be obtained from whole fruit [3], plum by-products [4], or juices/concentrates [5]. Commonly used drying industrial processes include, apart from conventional air-drying, freeze-drying (FD), vacuum drying (VD), and spray drying (SPD) [6]. However, the drying techniques, used to obtain powders from whole fruit, modify some of their physical and chemical properties [3,7]. In general, FD is considered one of the best method of obtaining high-quality products because the absence of liquid water and the low temperatures required for the process allows relatively high retention of bioactive compounds [8]. VD partially prevents thermal degradation of bioactive compounds in raw material because the temperature of the product is usually low and can be easily controlled [9]. And SPD is an effective technique in drying liquid products directly into powders, and is used broadly in processing dairy and fruit products, i.e., juices, extracts, or concentrates [10], and the resulting powders present a better preservation and retention of polyphenolic compounds [11,12]. Numerous studies confirmed that the above-mentioned processes had a strong influence on the physical properties of the dried whole fruits and pomace [3,5,13], as well as of the fruit juice powders [12].

Plums are phenolic-rich fruits that contain a mixture of polyphenolic compounds that can exert several biological effects, including antibacterial [14,15], antioxidant [5,16], and anti-inflammatory properties [17,18]. In commercially available plums, the most predominant and bioactive relevant compounds are phenolic acids, such as chlorogenic and neochlorogenic acids; flavonol glycosides, quercetin-3-glucoside and quercetin-3-galactoside; and anthocyanins, such as cyanidin and peonidin [19]. However, the profile and level of bioactive polyphenolic compounds in dried plum products is affected by the drying industrial processes. Until now, there is a lack of information in the literature on how the drying processes can modify the biological properties of fruit powders, regardless of the type of the fruit used for drying. Therefore, the objective of the present study was to evaluate the influence of different drying techniques applied for preparation of plum juice extract powders on the physical properties and the alterations in polyphenolic compounds contents and bioactive properties, i.e., antibacterial, antioxidant, and anti-inflammatory.

2. Materials and Methods

2.1. Reagents

3,4,5-Dimethylthiazol-2,5-diphenyl-tetrazolium bromide (MTT), carboxy-2′,7′-dichloro-dihydro-fluorescein diacetate (DCFH-DA), dimethyl sulfoxide (DMSO) were acquired from Sigma (Madrid, Spain). Dulbecco's Modified Eagle's Medium (DMEM), penicillin/streptomycin (5000 U/mL), phosphate buffered saline (PBS) and trypsin/EDTA solution (170,000 U/L) were purchased from Lonza (Madrid, Spain). Fetal bovine serum (FBS) of South American origin (Hyclone, GE Healthcare, Logan, UK) was obtained from Thermo Scientific (Madrid, Spain). Cell culture dishes were obtained from Sarstedt (Barcelona, Spain).

2.2. Materials

Material used in the study consisted of plum (*Prunus domestica* L.) (cv. Valor) juice obtained by laboratory hydraulic press (SRSE, Warsaw, Poland) that was used for preparation of polyphenolic extract. This was done by application of amberlite XAD-16 resin previously washed with water. The absorbed compounds were removed by ethanol that was evaporated by rotary evaporator Laborota 20 (Heidolph, Schwabag, Germany) at 40 °C to avoid the excessive degradation of the polyphenols. The solution was divided into 100 mL portions and subjected to the drying processes without the addition of any carrier agent due to the sugars' removal [5].

2.3. Drying Procedure

The plum juice extracts were submitted to drying techniques: (a) freeze-drying (FD) was performed in freeze-dryer (FreeZone, Labconco Corp., Kansas City, MO, USA) for 24 h (temperature of the chamber −60 °C and the heating plate +25 °C); (b) vacuum drying (VD) in Vacucell 111 Eco Line (MMM Medcenter Einrichtungen GmbH, Planegg, Germany) at the temperature of 40 °C, 60 °C and 80 °C at a pressure of 300 Pa for, respectively, 20, 16, and 10 h; (c) spray drying (SPD) done by mini spray dryer (Buchi, Flawil, Switzerland) with the inlet temperature of 180 °C and the outlet temperature of 70 °C. All drying techniques were performed at least in duplicate ($n = 2$). The obtained plum extracts powders (PEP) were milled (Bosch MKM 6003c, Gerlingen, Germany), vacuum packed (PP-5.14, Tepro SA, Koszalin, Poland), and stored at −20 °C until they were analyzed.

2.4. Physicochemical Properties

The moisture content of PEP was determined in duplicate ($n = 2$) according to Figiel et al. [20] using Vacucell 111 Eco Line at 80 °C for 20 h. The results were expressed as %. The water activity of the PEP was done at 25 °C by Water Activity Meter AquaLab (DewPoint 4Te, Decagon Devices Inc., Pullman, WA, USA) ($n = 2$). The colour of the samples was determined in a CIE $L^*a^*b^*$ system using a Minolta Chroma Meter CR-400 (Minolta Co. Ltd., Osaka, Japan). Data were presented as an average value of five measurements ($n = 5$).

2.5. Characterization of Polyphenolic Compounds by UPLC-PDA System

The preparation of the PEP extracts subjected to characterization of polyphenolic compounds using the UPLC-PDA system was performed according to Wojdyło et al. [21]. All determinations were done in duplicate. The results were expressed as grams per 100 g of dry basis (db) of plum extract power (g/100 g db).

2.6. Bacterial Strains, Growth Media, and Culture Conditions

Five of the most relevant foodborne pathogen bacteria were tested for antibacterial activity of PEP: *Campylobacter jejuni* NCTC11168, *Escherichia coli* ATCC®25922™, *Staphylococcus aureus* ATCC®25923™, *Listeria monocytogenes* CECT935, and *Salmonella enterica* subsp. enterica serovar Typhimurium ATCC® 14028™. All bacteria strains were stored at −80 °C in Brucella Broth (BB) (Becton, Dickinson, & Co, Madrid, Spain) plus 20% glycerol. The agar-plating medium consisted of Müeller-Hinton agar supplemented with 5% defibrinated sheep blood (MHB) (Becton, Dickinson, & Co), and liquid growth medium consisted of BB. Bacteria cultures were prepared as follows: The frozen stored strains were reactivated by inoculation in MHB and incubation under aerobic conditions (*E. coli*, *S. aureus*, *L. monocytogenes*, and *S. typhimurium*) and microaerophilic conditions (*C. jejuni*) using a Variable Atmosphere Incubator (85% N_2, 10% CO_2, 5% O_2) (VAIN, MACS-VA500, Don Whitley Scientific, Bingley, UK) at 37 °C for 24–48 h. Isolated colonies were inoculated into 50 mL of BB and incubated in the conditions described above following stirring (150 rpm) for 24–48 h until the late exponential phase, and used as experimental inoculum. These bacterial inoculum cultures ~1 × 108 colony forming units (CFU/mL) were used for the different experimental assays.

2.7. Antibacterial Activity

The antibacterial activity of PEP against foodborne bacteria was evaluated following the procedure described by Silvan et al. [22]. Briefly, 1 mL of PEP (1 mg/mL final concentrations) was transferred into different flasks containing 4 mL of BB. Bacterial inoculum (50 µL of ~1×10^8 CFU/mL) was then inoculated into the flasks under aseptic conditions. The culture was prepared in triplicate and incubated under stirring (150 rpm) during 24 h at 37 °C. Growth controls were prepared by transferring 1 mL of sterile water to 4 mL of BB and 50 µL of bacterial inoculum. After incubation, serial decimal dilutions of mixtures were prepared in saline solution (0.9% NaCl) and they were plated (20 µL) onto fresh MHB agar and incubated as previously described. The number of CFU was assessed after incubation. The results of antibacterial activity were expressed as percentage of growth inhibition for the foodborne bacteria respect to the controls of bacteria growth.

2.8. Cell Cultures

Human intestinal epithelial HT-29 and murine macrophage RAW 264.7 cell lines were obtained from the American Type Culture Collection (ATCC). Both cell lines were cultured in Dulbecco's modified Eagles's medium (Lonza) supplemented with 10% fetal bovine serum (Hyclone) and 1% penicillin/streptomycin (5000 U/mL, Lonza). The cells were plated at densities of ~1×10^6 cells in 75 cm^2 culture flasks and maintained at 37 °C under 5% CO_2 in a humidified incubator until 90% confluence. The culture medium was changed every 2 days. Before a confluent monolayer appeared, sub-culturing cell process was carried out.

2.9. Cell Viability

Before the cellular antioxidant and anti-inflammatory experiments, it was necessary to find out if the plum powders were cytotoxic for both cell lines (HT-29 and RAW 264.7). With this purpose, cell viability was determined by MTT reduction assay as previously was described by Silvan et al. [23]. Confluent stock cultures (~90%) were trypsinized (Trypsin/EDTA), and cells were seeded in 96-well plates (~5×10^4 cells per well) and incubated in culture medium at 37 °C under 5% CO_2 in a humidifier incubator for 24 h. Briefly, cell medium was replaced with serum-free medium containing PEP (1 mg/mL final concentrations), and the cells were incubated at 37 °C for 24 h under 5% CO_2. Control cells (non-treated) were incubated in serum-free medium without plum powders. Thereafter, cells were washed with PBS, the medium was replaced by 200 μL of serum-free medium, and 20 μL of MTT solution in PBS (5 mg/mL) was added to each well for the quantification of the living metabolically active cells after 1-h incubation. MTT is reduced to purple formazan in the mitochondria of living cells. Formazan crystals in the wells were solubilized in 200 μL DMSO. Finally, the absorbance was measured at 570 nm wavelength by employing a microplate reader Synergy HT (BioTek Instruments Inc., Winooski, VT, USA). The viability was calculated considering controls containing serum-free medium as 100% viable. Data represent the mean and standard deviation of three independent experiments (n = 3). All experiments were carried out between passage 10 to passage 30 to ensure cell uniformity and reproducibility.

2.10. Antioxidant Activity Against Intracellular Reactive Oxygen Species (ROS) Production

Human intestinal epithelial cell line HT-29 was used for the evaluation of oxidative stress. Intracellular ROS were measured by the DCFH-DA assay as previously was reported by Martín et al. [24]. Cells were seeded (5×10^4 cells per well) in 24-well plates and grown until they reached 70% of confluence. Cells were pre-treated with plum powders (1 mg/mL) dissolved in serum-free medium for 24 h. After that, the cells were washed with PBS and incubated with 20 μM DCFH-DA for 30 min at 37 °C. Then, cells were washed twice with PBS to remove the unabsorbed probe and treated with serum-free medium, containing 2.5 mM tert-butyl hydroperoxide (TBHP). ROS production was immediately monitored for 180 min in a fluorescent microplate reader Synergy HT using a λ_{ex} 485 nm and λ_{em} 530 nm. After being oxidized by intracellular oxidants, DCFH-DA changes to dichlorofluorescein (DCF) and emits fluorescence. The cells treated with TBHP was used as oxidative control (100% of intracellular ROS production). All samples were analyzed in triplicate (n = 3). All experiments were carried out between passage 10 to passage 30 to ensure cell uniformity and reproducibility. Data were expressed as percentage of fluorescence generation relative to negative control cells.

2.11. Anti-Inflammatory Activity

For inflammatory experiments, murine macrophage cell line RAW 264.7 was used. Cells were seeded in a 96-well plate at density of ~5 × 104 cells/well. Nitrite accumulation, indicator of nitric oxide (NO) synthesis, was measured in the culture medium of treated and control cells by the Griess reaction. After 24 h of incubation, the medium was removed, the cells were washed with 200 μL PBS and treated with PEP (1 mg/mL) and 10 μg/mL of LPS from *Escherichia coli* O55:B5 for 24 h. Control cells were incubated in serum-free medium with LPS (Control+) and without LPS (Control−) for 24 h. Finally, the media were collected and used for NO quantification. Briefly, 100 μL of collected cell supernatants were plated in 96-well plate and an equal amount of Griess reagent constituted by 1% (*w*/*v*) sulfanilamide and 0.1% (*w*/*v*) N-1-(naphthyl) ethylenediamine-diHCl in 2.5% (*v*/*v*) H_3PO_4, was added. The plate was incubated for 5 min and the absorbance measured at 550 nm in a microplate reader Synergy HT. The amount of NO was calculated using a sodium nitrite standard curve (0–10 μg/mL). Data were expressed as percentage of NO production calculated relative to Control+. Data represent the mean and standard deviation of three independent experiments ($n = 3$). All experiments were carried out between passage 10 to passage 30 to ensure cell uniformity and reproducibility.

2.12. Statistical Analysis

The results were reported as means ± standard deviations (SD) performed in triplicate. A *t*-test was used to assess the differences in antibacterial activity. Significant differences among the data were estimated by applying analysis of variance (ANOVA). The Tukey's least significant differences (LSD) test was used to evaluate the significance of the analysis. Differences were considered significant at $p < 0.05$. All statistical tests were performed with IBM SPSS Statistics for Windows, Version 25.0 (IBM Corp., Armonk, NY, USA).

3. Results and Discussion

3.1. Physical Properties

The powders obtained from plum juice extracts submitted to different drying techniques differed in terms of moisture content and water activity (Table 1). The moisture content ranged from 2.44% up to 7.34% and was within the range described by Boonyai et al. [25]. The values of moisture content depended on the drying method and the parameters applied. The lowest water content was noted after spray drying during which the highest temperature for water removal was applied [12]. The evaluation of water activity in powders is a key aspect as this parameter informs about their stability, both chemical and microbial. It is connected with the quality of the dried products as the rate for some chemical reactions begins above a water activity of 0.3 [26]. Similarly, a high positive correlation ($r = 0.789$) between moisture content and water activity have been previously described in fruit powders [12]. PEP also differed in color attributes (Table 1).

Table 1. Physical properties of the plum extracts powders (PEP) obtained after different drying techniques.

	Moisture Content (%)	Water Activity (-)	Colour				
			L^*	a^*	b^*	C^*	h^*
FD	7.34 ± 0.14 d	0.289 ± 0.001 d	42.84 ± 0.13 b	23.37 ± 0.1 d	10.73 ± 0.03 d	25.72 ± 0.10 c	24.67 ± 0.13 b
VD 40 °C	3.74 ± 0.17 c	0.227 ± 0.002 b	40.28 ± 0.05 a	17.73 ± 0.04 c	7.64 ± 0.01 b	19.31 ± 0.03 b	23.33 ± 0.05 a
VD 60 °C	7.17 ± 0.04 d	0.243 ± 0.001 c	43.10 ± 0.67 b	16.94 ± 0.61 b	8.33 ± 0.10 c	18.88 ± 0.59 b	26.19 ± 0.56 c
VD 80 °C	3.29 ± 0.08 b	0.242 ± 0.001 c	40.40 ± 0.09 a	13.56 ± 0.02 a	6.64 ± 0.02 a	15.11 ± 0.03 a	26.09 ± 0.07 c
SPD	2.44 ± 0.06 a	0.146 ± 0.002 a	45.76 ± 0.02 c	24.97 ± 0.06 e	13.15 ± 0.03 e	28.22 ± 0.04 d	27.78 ± 0.09 d

Freeze drying (FD); Vacuum drying 40 °C (VD 40 °C), 60 °C (VD 60 °C), 80 °C (VD 80 °C); Spray drying (SPD); colour parameters (L^*, a^*, b^*), chroma (C^*) and hue (h). Different letters (a,b,c,d,e) within the same column indicated statistical differences between samples ($p \leq 0.05$; LSD Tukey).

Among products obtained, lower values of coordinate L^* were noted after VD, when compared to FD and SPD, that was in agreement with previously conducted research on chokeberry [11] and apple powders [12]. The L^* values were connected with coordinate a* and b* pointing a strong influence of the drying technique on the retention of red and yellow pigments. Chroma (C^*) is connected with the color intensity and opposite to apple powders [12], the highest values were linked to FD and SPD processes. This could be connected with the fact that, in the current study, the extract of plum juice was dried, and thereby the material differed in terms of the chemical composition, as the obtained extracts were significantly darker when compared to the juice. Taking the above into consideration, the chemical changes that have occurred during drying are strictly connected with the chemical properties of the initial materials. The hue angle (h^*) values indicated that the analyzed samples were more reddish as an angle of approximately 0 represents red colour. The highest h^* values were obtained for powders gained after FD that was in line with the coordinate a^*. Additionally, the lowest values were noted after FD and VD 40 °C, pointing to a strong influence on the temperature of the process on this parameter.

3.2. Polyphenolic Compounds Composition

The total polyphenolic compounds content in PEP ranged from 34.66 to 47.65 g/100 g db and differed due to the drying technique and parameters applied for their dehydration (Table 2) [5]. In general, the highest total content of identified polyphenolic compounds was noted after SPD, which was almost 19% higher when compared to FD, and is regarded as the technique that preserve bioactive compounds the most. Similarly, the highest retention of polyphenolic compounds in sugar-free extracts, submitted to SPD, was noted in case of chokeberry [11] and cranberry [27], which pointed to this dehydration method as being successfully used when the retention of selected polyphenolic compounds is concerned.

Table 2. Content of identified major phenolic compounds in plum extracts powders (PEP) obtained by selected drying methods (g/100 g db).

Compound	FD	VD 40 °C	VD 60 °C	VD 80 °C	SPD
		Phenolic acids			
Neochlorogenic acid	10.25 ± 0.12 c	10.18 ± 0.02 c	8.85 ± 0.11 b	6.45 ± 0.02 a	11.04 ± 0.33 c
3-feruoylquinic acid	10.51 ± 0.06 c	9.96 ± 0.11 c	8.51 ± 0.36 b	5.42 ± 0.15 a	11.69 ± 0.28 d
3-O-p-coumaroylquinic acid	4.41 ± 0.09 ab	4.73 ± 0.11 b	4.59 ± 0.15 ab	4.05 ± 0.21 a	5.30 ± 0.14 c
Chlorogenic acid	4.14 ± 0.01 c	3.96 ± 0.01 bc	3.58 ± 0.02 ab	3.36 ± 0.12 a	4.70 ± 0.04 d
Methyl 3-caffeoylquinate	0.25 ± 0.01 a	0.37 ± 0.01 ab	0.55 ± 0.02 b	7.08 ± 0.11 d	3.92 ± 0.06 c
Total phenolic acids	29.56	29.20	26.08	26.36	36.65
		Flavonols			
Quercetin-3-O-rutinoside	1.25 ± 0.01 c	0.09 ± 0.01 a	0.24 ± 0.04 b	1.09 ± 0.16 c	1.88 ± 0.17 d
Quercetin-3-O-galactoside	5.01 ± 0.05 d	1.29 ± 0.01 b	1.41 ± 0.01 c	1.47 ± 0.06 c	0.73 ± 0.04 a
Quercetin-3-O-glucoside	0.73 ± 0.01 a	5.24 ± 0.02 c	5.14 ± 0.07 c	4.58 ± 0.09 b	6.11 ± 0.14 d
Quercetin-3-O-(6″ acetylgalactoside)	1.03 ± 0.01 c	1.04 ± 0.01 c	0.89 ± 0.11 b	0.61 ± 0.01 a	1.21 ± 0.09 d
Total flavonols	8.02	7.66	7.68	7.75	9.93
		Anthocyanins			
Cyanidin-3-O-glucoside	0.27 ± 0.02 c	0.26 ± 0.01 c	0.24 ± 0.04 b	0.17 ± 0.03 a	0.28 ± 0.01 c
Cyanidin-3-O-rutinoside	0.70 ± 0.01 bc	0.70 ± 0.00 bc	0.66 ± 0.11 ab	0.53 ± 0.03 a	0.78 ± 0.04 bc
Peonidin-3-O-rutinoside	0.0058 ± 0.0001 b	0.0058 ± 0.0002 b	0.0055 ± 0.0009 b	0.0044 ± 0.0007 a	0.0065 ± 0.0003 c
Total anthocyanins	0.97	0.96	0.90	0.70	1.07
Total polyphenolics content	38.55	37.82	34.66	34.81	47.65

Freeze drying (FD); Vacuum drying 40 °C (VD 40 °C), 60 °C (VD 60 °C), 80 °C (VD 80 °C); Spray drying (SPD); different letters (a,b,c,d) within the rows indicate significant differences between samples ($p \leq 0.05$; LSD Tukey test).

Going into the details, the major group of polyphenolic compounds present in PEP consisted of five identified phenolic acids, among which the chlorogenic and 3-feruoylquinic acids were dominant. The highest retention of these constituents was noted after SPD. As previously observed [5], the drying processes led to the degradation of phenolic acids within the increase of the temperature during VD, except methyl 3-caffeoylquinate. Interestingly, the increase in the temperature during VD up to 80 °C, followed by SPD (170 °C) resulted in, respectively, 29-, and 16-times higher content of this compound in the powders when compared to those gained after FD. In case of flavonols, processing during which the temperature above 40 °C (VD 40 °C, VD 60 °C, VD 80 °C and SPD) caused a strong degradation of quercetin-3-O-galactoside, the application of such conditions caused an increase in the content of quercetin-3-O-glucoside when compared to FD. This might be connected with the thermally-induced release of this constituent during drying. In general, the SPD process resulted in better retention of flavonols being almost 20% higher when compared to FD. In the current study, three anthocyanins had been identified and their content followed the same patterns as the above-mentioned groups of polyphenolic compounds.

3.3. Antibacterial Activity

The antibacterial activity of PEP against five representative foodborne pathogens (*C. jejuni*, *S. typhimurium*, *E. coli*, *S. aureus*, and *L. monocytogenes*) was evaluated. As shown in Figure 1, all extracts were active against at least one of the pathogens studied, and their antibacterial activity was related to the drying procedure used for their conservation. PEP exhibited different levels of growth inhibition against the foodborne pathogens evidencing a strain-dependent effect. Among all the studied powders, the most relevant antibacterial activity was observed for FD extract. In fact, this extract inhibited significantly ($p < 0.05$) the growth of all the bacteria studied except *E. coli* strain. This extract showed a growth inhibition range between 22–52%, depending on the bacterial strain. VD 60 °C extract inhibited the growth of three of the five foodborne bacterial strains (*C. jejuni*, *S. aureus*, and *L. monocytogenes*), compressing a growth inhibition range of 17–36%. VD 40 °C and VD 80 °C inhibited the growth of two bacterial strains (*C. jejuni-L. monocytogenes*, and *E. coli-L. monocytogenes*) with an inhibition range of 46–58%, and 26–46%, respectively. Otherwise, SPD extract inhibited only the growth of *L. monocytogenes* strain (17% of inhibition). Taking into account the nature of the microorganism, the results showed that *L. monocytogenes* was inhibited by all the PEP in the range of 17%–46%, regardless of the drying process used (Figure 1).

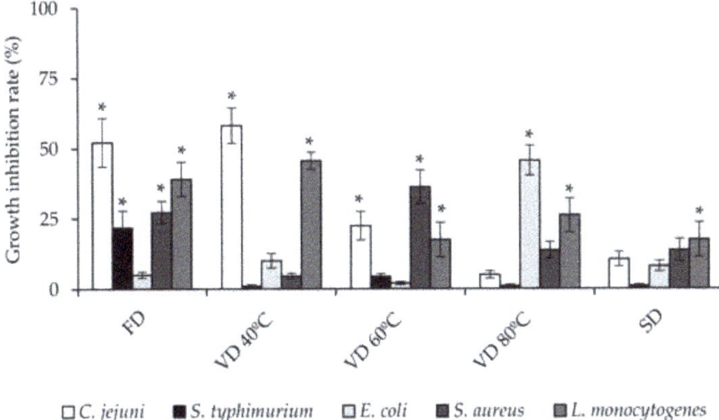

Figure 1. Effect of PEP (1 mg/mL) on foodborne bacteria growth after 24 h of incubation. Results represent the percentage of growth inhibition respect to the untreated control (100% of growth) and are expressed as mean ±SD ($n = 3$). Bars marked with asterisk indicate significant growth inhibition respect to the control by *t*-test ($p \leq 0.05$). Freeze drying (FD); Vacuum drying 40 °C (VD 40 °C), 60 °C (VD 60 °C), 80 °C (VD 80 °C); Spray drying (SPD).

C. jejuni, the leading cause of bacterial foodborne diarrheal illness worldwide, was also inhibited for three of the powders used (FD, VD 40 °C, and VD 60 °C) in a range of 22–58%. FD and VD 60 °C (27–36%) affected the growth of *S. aureus*, while *S. typhimurium* and *E. coli* were the bacterial strains with the lowest sensitivity to the all PEP. The different antibacterial activities by PEP against foodborne bacterial strains may be related to the different composition of phenolic compounds of each sample (Table 2), assuming that it is generally accepted that phenolic compounds, present in plant extracts, play a mandatory role in their antibacterial effects [28]. However, as can be deduced from Table 2, it is not the total concentration of phenolic compounds present in the sample, which determines its antibacterial effect, but rather the presence of certain specific polyphenolic compounds in the extract. In this regard, FD was the most active bactericidal extract (Figure 1), showed a significant higher concentration of quercetin-3-*O*-galactoside (hyperoside). Hyperoside is a flavonol glycoside with variety of biological activities, including anti-inflammatory, antioxidant, and antimicrobial activities [29–31]. Its antibacterial effect has been demonstrated both, against gram-negative bacteria such as *P. aeruginosa* [32] and against gram-positive bacteria, such as *S. aureus* [33]. The results obtained in this work suggest that the hyperoside could be involved in the antimicrobial effect observed, since this compound has a significantly higher concentration in the powder obtained by FD, while the rest of the phenolic compounds identified are in concentrations similar or lower than those obtained for the rest of the extracts (Table 2). Apparently, the antibacterial effect of PEP used to be more effective against gram-positive bacteria [14]. This behavior is influenced by differences in the cell membrane constituents. Gram-positive bacteria contain an outer peptidoglycan layer, which is an ineffective permeability barrier; meanwhile gram-negative bacteria have outer phospholipidic membrane carrying structural lipopolysaccharide components, which represent an obstacle for polyphenolic compounds to enter the cell cytoplasm [14,34]. This pattern is also observed in our work, except for *C. jejuni*. Although it is a gram-negative bacterium, *Campylobacter* lacks many of the genetic regulatory networks found in other gram-negative bacteria that allow them to respond to, and cope with, adverse conditions [35]. Accordingly, we have previously demonstrated that *Campylobacter* can be significantly inhibited by different polyphenolic compounds [22,36]. Therefore, antibacterial activity of PEP could be modulated depending of the drying procedure used. The drying process involves several variables, which can change the polyphenolic composition of the extract, resulting in a modified antibacterial response.

3.4. Antioxidant Activity Against Intracellular Reactive Oxygen Species (ROS) Production

Oxidative stress is involved in several acute and chronic pathological processes due to its ability in activating inflammatory pathways [37,38]. The antioxidant activity of plum polyphenolic compounds has been previously investigated in different system models, such as ABTS, DPPH, FRAP, and ORAC assays [5,16]. However, due to the absence of information on the scavenger activity of ROS in biological models, we investigated the ability of PEP in reducing intracellular oxidative stress. The experiment was carried out using similar concentrations of PEP to those used in the antibacterial activity assay (1 mg/mL). This concentration did not significantly affect cell viability at 24 h after treatment (data not shown). As shown in Figure 2, HT-29 cells, exposed to TBHP, increased the ROS level (oxidative control). Pre-treatment of intestinal cells with PEP showed a significant ($p < 0.05$) reduction in cellular ROS generation, stimulated by TBHP, compared with the oxidative control, except when the cells were pre-treated with VD 40 °C sample. VD 80 °C and SPD caused the highest protection against oxidative damage in stressed cells, inducing an inhibition percentage of ROS production close to 37% respect to the oxidative control cells. Although, FD and VD 60 °C had a significant antioxidant activity with respect to the experimental control, this was lower than the samples obtained with treatments carried out at higher temperatures. Previously, others have shown that drying processes can influence the antioxidant capacity of the dried products, observing an increase in the antioxidant capacity of plum extracts dried by microwave vacuum at high temperatures [3]. Apparently, the temperature applied during plum juice drying is related to the increase in the content of polyphenolic compounds able to scavenge ROS, as the highest values of those compounds were noted after drying at 80 °C. In this regard, the drying temperature of the PEP and the subsequent polyphenolic composition affected the antioxidant capacity of the obtained powders, therefore, the extracts exposed to higher temperatures during the drying process showed a higher content in the methyl 3-caffeoylquinate (Table 2) and significantly improved antioxidant activities. It is well known that phenolic acids can prevent oxidative damage because of their ability to scavenge ROS [39]. Within the phenolic acids, chlorogenic acid and their derivatives, such as methyl 3-caffeoylquinate acid, act as potent ROS scavengers by donating hydrogen atoms to reactive molecules, transforming them to less active radicals, and maintaining an optimal cellular oxidative balance [40–42].

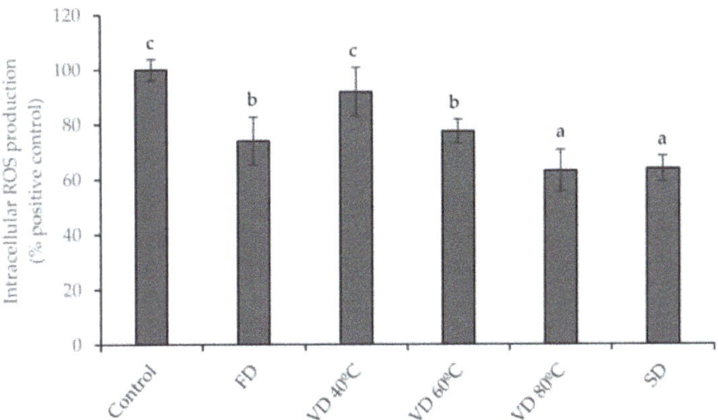

Figure 2. Protective effect of PEP (1 mg/mL) on intracellular ROS production. HT-29 cells were incubated with the powders for 24 h and then treated with 2.5 mM TBHP for 3 h, and ROS production was determined. Values are expressed as a percentage relative to the control conditions and are represented by mean ±SD (n = 3). Bars with different letters indicate significant differences on ROS production by ANOVA post hoc LSD Tukey test ($p \leq 0.05$). Freeze drying (FD); Vacuum drying 40 °C (VD 40 °C), 60 °C (VD 60 °C), 80 °C (VD 80 °C); Spray drying (SPD).

Our findings show that PEP could induced cellular protection against TBHP-induced oxidative stress by preventing intracellular ROS accumulation. However, the level of antioxidant capacity of the PEP may be determined by the conditions applied during the drying process.

3.5. Anti-Inflammatory Activity

We investigated the impact of PEP on the inflammatory process in RAW264.7 cells stimulated with LPS. During an inflammatory event, the inducible enzyme nitric oxide synthase (iNOS) is responsible for an exacerbated production of NO, which can lead to tissue lesions, organ dysfunction, and inflammation-related diseases [43]. As shown in Figure 3, LPS significantly ($p < 0.05$) stimulated the production of NO in macrophages (Control+) with respect to non-stimulated cells (Control−). The percentage of NO production was significantly ($p < 0.05$) reduced by 24–39% in RAW264.7 cells pre-treated with PEP. Pre-treatment of LPS-stimulated cells with VD 80 °C and VD 60 °C powders led to greater attenuation of NO production with an inhibition range between 33% and 39%. The rest of the extracts had a similar behavior ($p < 0.05$), regardless of the drying treatment conducted, and caused an attenuation in the production of NO in a range between 24% and 30%. Apparently, a greater anti-inflammatory capacity would be influenced both, by the type of treatment used (VD > SPD > FD) and by the temperature used in each treatment. Overall, this behavior is similar to that observed in the results obtained from the analysis of antioxidant activity (ROS inhibition), where VD 80 °C was the most active sample in both determinations. Some investigators have demonstrated that the cellular inflammatory responses are caused because of ROS production [17,44]. Consequently, a reduction in intracellular ROS levels could lead to an inhibition of the inflammation process through the reduction of NO production. This postulate agrees with our results, since the samples that showed high antioxidant activity by reducing intracellular ROS (Figure 2) also showed relevant anti-inflammatory properties by reducing NO production (Figure 3).

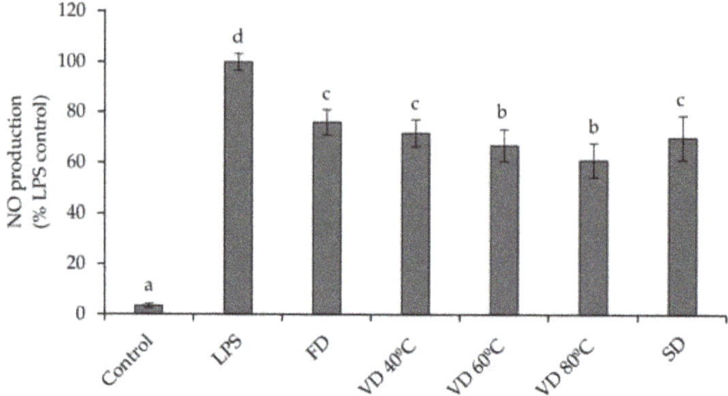

Figure 3. Effect of PEP (1 mg/mL) on nitric oxide (NO) production in LPS-stimulated RAW264.7 macrophage cells. Values are expressed as a percentage relative to the LPS-stimulated control group and are represented by mean ±SD (n = 3). Bars with different letters indicate significant differences on NO production by ANOVA LSD Tukey test ($p \leq 0.05$). Freeze drying (FD); Vacuum drying 40 °C (VD 40 °C), 60 °C (VD 60 °C), 80 °C (VD 80 °C); Spray drying (SPD).

However, it is difficult to precise the main polyphenolic compounds that would be most involved in the anti-inflammatory response observed, although apparently, several of them appear to be involved. Previous studies have demonstrated that the anti-inflammatory effects of plum phenolic compounds [1,18], mainly due to the presence of phenolic acids, can decrease the expression of inflammatory mediators, such as nuclear factor κB (NF-κB), vascular cell adhesion molecule 1 (VCAM-1), cyclooxygenase-2 (COX-2), and iNOS mRNA [45]. In addition, plum phenolic compounds can contribute to the modulation of the inflammatory responses in human cells by inhibiting various inflammatory factors, such as cytokines IL-6 and IL-8 [17]. In addition, the potential anti-inflammatory activity of the PEP could be particularly important, considering that activation inflammatory pathways can stimulate proliferation of cancer cells [38].

4. Conclusions

This study has shown that plum extract powders gained after freeze-, vacuum-, and spray-drying have promising antibacterial, antioxidant, and anti-inflammatory properties that have been tested in different biological models. The drying processes significantly influences both, the physical properties and the composition of polyphenols, and thus, the bioactive properties plum juice extract powders. The drying techniques moderated the content of polyphenolic compounds in the powders, which influence the different levels of growth inhibition against the foodborne pathogens, evidencing a strain-dependent effect being the most relevant for FD extract. This extract inhibited significantly ($p < 0.05$) the growth of all the bacteria studied, except *E. coli* strain. It was observed that *L. monocytogenes* was inhibited by all the dried plum juice extracts in the range of 17–46%, regardless of the drying process used. These powders could also induce cellular protection against oxidative stress by preventing intracellular ROS accumulation, but the level of antioxidant capacity may be determined by the conditions applied during the drying process. It was shown that VD 80 °C and SPD caused the highest protection against oxidative damage in stressed cells, inducing an inhibition percentage of ROS production close to 37% respect to the oxidative control cells. Moreover, plum extracts powders exhibited a greater anti-inflammatory capacity, which would be influenced both, by the type of treatment used and by the temperature used in each treatment, being the VD 80 °C - the sample with the highest protection level. The results demonstrate that the drying method selected can be an effective tool for modulating the composition, physical, and bioactive properties of plum extracts powders.

Author Contributions: Conceptualization, J.M.S., A.M.-C., and A.J.M.-R.; data curation, A.M.-C., and A.J.M.-R.; Formal analysis, J.M.S. and A.M.-C.; funding acquisition, A.J.M.-R.; investigation, J.M.S., A.M.-C., and A.J.M.-R.; Methodology, J.M.S. and A.M.-C.; project administration, A.J.M.-R.; resources, A.J.M.-R.; software, J.M.S., and A.M.-C.; supervision, A.J.M.-R.; writing-original draft, J.M.S., A.M.-C., and A.J.M.-R.; writing-review and editing, J.M.S., A.M.-C., and A.J.M.-R. All authors have read and agreed to the published version of the manuscript.

Funding: This research was funded by AGL2017-89566-R project from the Consejo Superior de Investigaciones Científicas (Spain).

Acknowledgments: This work was founded through Project AGL2017-89566-R from the Consejo Superior de Investigaciones Científicas (Spain). Authors are grateful to Krzysztof Lech and Adam Figiel from Institute of Agricultural Engineering (Wroclaw University of Environmental and Life Sciences) for possibility to perform water activity and color measurements.

Conflicts of Interest: The authors declare no conflict of interest. The funders had no role in the design of the study; in the collection, analyses, or interpretation of data; in the writing of the manuscript, or in the decision to publish the results.

References

1. Noratto, G.; Porter, W.; Byrne, D.; Cisneros-Zevallos, L. Identifying peach and plum polyphenols with chemopreventive potential against estrogen-independent breast cancer cells. *J. Agric. Food Chem.* **2009**, *57*, 5219–5226. [CrossRef] [PubMed]
2. FAOSTAT. Food and Agriculture Organization Corporate Statistical Database. Production. Available online: http://www.fao.org/faostat/en/#data/QC/visualize (accessed on 20 November 2019).
3. Michalska, A.; Wojdyło, A.; Lech, K.; Łysiak, G.P.; Figiel, A. Physicochemical properties of whole fruit plum powders obtained using different drying technologies. *Food Chem.* **2016**, *207*, 223–232. [CrossRef] [PubMed]
4. Michalska, A.; Wojdyło, A.; Majerska, J.; Lech, K.; Brzezowska, J. Qualitative and quantitative evaluation of heat-induced changes in polyphenols and antioxidant capacity in *Prunus domestica* L. by-products. *Molecules* **2019**, *24*, 3008. [CrossRef] [PubMed]
5. Michalska, A.; Wojdyło, A.; Łysiak, G.P.; Figiel, A. Chemical composition and antioxidant properties of powders obtained from different plum juice formulations. *Int. J. Mol. Sci.* **2017**, *18*, 176. [CrossRef] [PubMed]
6. Figiel, A.; Michalska, A. Overall quality of fruits and vegetables products affected by the drying processes with the assistance of vacuum-microwaves. *Int. J. Mol. Sci.* **2017**, *18*, 71. [CrossRef] [PubMed]
7. Karam, M.C.; Petit, J.; Zimmer, D.; Djantou, E.B.; Scher, J. Effects of drying and grinding in production of fruit and vegetable powders: A review. *J. Food Eng.* **2016**, *188*, 32–49. [CrossRef]
8. Ratti, C. Hot air and freeze-drying of high-value foods: A review. *J. Food Eng.* **2001**, *49*, 311–319. [CrossRef]
9. Lewicki, P.P. Design of hot air drying for better foods. *Trends Food Sci. Technol.* **2006**, *17*, 153–163. [CrossRef]
10. Alamilla-Beltrán, L.; Chanona-Pérez, J.J.; Jiménez-Aparicio, A.R.; Gutiérrez-López, G.F. Description of morphological changes of particles along spray drying. *J. Food Eng.* **2005**, *67*, 179–184. [CrossRef]
11. Horszwald, A.; Julien, H.; Andlauer, W. Characterisation of Aronia powders obtained by different drying processes. *Food Chem.* **2013**, *141*, 2858–2863. [CrossRef]
12. Michalska, A.; Lech, K. The effect of carrier quantity and drying method on the physical properties of apple juice powders. *Beverages* **2018**, *4*, 2. [CrossRef]
13. Majerska, J.; Michalska, A.; Figiel, A. A review of new directions in managing fruit and vegetable processing by-products. *Trends Food Sci. Technol.* **2019**, *88*, 207–219. [CrossRef]
14. Sójka, M.; Kołodziejczyk, K.; Milala, J.; Abadias, M.; Viñas, I.; Guyot, S.; Baron, A. Composition and properties of the polyphenolic extracts obtained from industrial plum pomaces. *J. Funct. Foods* **2015**, *12*, 168–178. [CrossRef]
15. Valtierra-Rodriguez, D.; Heredia, N.L.; Garcia, S.; Sanchez, E. Reduction of Campylobacter jejuni and Campylobacter coli in poultry skin by fruit extracts. *J. Food Prot.* **2010**, *73*, 477–482. [CrossRef]
16. Khallouki, F.; Haubner, R.; Erben, G.; Ulrich, C.M.; Owen, R.W. Phytochemical composition and antioxidant capacity of various botanical parts of the fruits of *Prunus domestica* L. from the Lorraine region of Europe. *Food Chem.* **2012**, *133*, 697–706. [CrossRef]
17. Kaulmann, A.; Legay, S.; Schneider, Y.J.; Hoffmann, L.; Bohn, T. Inflammation related responses of intestinal cells to plum and cabbage digesta with differential carotenoid and polyphenol profiles following simulated gastrointestinal digestion. *Mol. Nutr. Food Res.* **2016**, *60*, 992–1005. [CrossRef]

18. Hooshmand, S.; Kumar, A.; Zhang, J.Y.; Johnson, S.A.; Chaid, S.C.; Arjmandi, B.H. Evidence for anti-inflammatory and antioxidative properties of dried plum polyphenols in macrophage RAW 264.7 cells. *Food Funct.* **2015**, *6*, 1719–1725. [CrossRef]
19. Treutter, D.; Wang, D.; Farag, M.A.; Argueta Baires, G.D.; Ruhmann, S.; Neumuller, M. Diversity of phenolic profiles in the fruit skin of Prunus domestica plums and related species. *J. Agric. Food Chem.* **2012**, *60*, 12011–12019. [CrossRef]
20. Figiel, A. Drying kinetics and quality of beetroots dehydrated by combination of convective and vacuum-microwave methods. *J. Food Eng.* **2010**, *98*, 461–470. [CrossRef]
21. Wojdyło, A.; Oszmianski, J.; Bielicki, P. Polyphenolic composition, antioxidant activity, and polyphenoloxidase (PPO) activity of quince (*Cydonia oblonga Miller*) varieties. *J. Agric. Food Chem.* **2013**, *61*, 2762–2772. [CrossRef]
22. Silvan, J.M.; Mingo, E.; Hidalgo, M.; de Pascual-Teresa, S.; Carrascosa, A.V.; Martinez-Rodriguez, A.J. Antibacterial activity of a grape seed extract and its fractions against Campylobacter spp. *Food Control* **2013**, *29*, 25–31. [CrossRef]
23. Silvan, J.M.; Zorraquin-Peña, I.; Gonzalez de Llano, D.; Moreno-Arribas, M.V.; Martinez-Rodriguez, A.J. Antibacterial activity of glutathione-stabilized silver nanoparticles against Campylobacter multidrug-resistant strains. *Front. Microbiol.* **2018**, *9*, 458. [CrossRef] [PubMed]
24. Martín, M.A.; Cordero-Herrera, I.; Bravo, L.; Ramos, S.; Goya, L. Cocoa flavanols show beneficial effects in cultured pancreatic beta cells and liver cells to prevent the onset of type 2 diabetes. *Food Res. Int.* **2014**, *63*, 400–408. [CrossRef]
25. Boonyai, P.; Howes, T.; Bhandari, B. Applications of the cyclone stickiness test for characterization of stickiness in food powders. *Dry. Technol.* **2006**, *24*, 703–709. [CrossRef]
26. Rahman, M.S.; Labuza, T.P. Water activity and food preservation. In *Handbook of Food Preservation*; Rahman, S., Ed.; CRC Press, Taylor & Francis Group: Boca Raton, FL, USA, 2007; pp. 447–476.
27. Michalska, A.; Wojdyło, A.; Honke, J.; Ciska, E.; Andlauer, W. Drying-induced physico-chemical changes in cranberry products. *Food Chem.* **2018**, *240*, 448–455. [CrossRef]
28. Martillanes, S.; Rocha-Pimienta, J.; Cabrera-Bañegil, M.; Martín-Vertedor, D.; Delgado-Adámez, J. Application of phenolic compounds for food preservation: Food additive and active packaging. In *Phenolic Compounds Biological Activity*; Soto-Hernández, M., Ed.; IntechOpen: London, UK, 2017; pp. 39–57.
29. Ku, S.K.; Kwak, S.; Kwon, O.J.; Bae, J.S. Hyperoside inhibits high-glucose-induced vascular inflammation in vitro and in vivo. *Inflammation* **2014**, *37*, 1389–1400. [CrossRef]
30. Xing, H.Y.; Liu, Y.; Chen, J.H.; Sun, F.J.; Shi, H.Q.; Xia, P.Y. Hyperoside attenuates hydrogen peroxide-induced L02 cell damage via MAPK-dependent Keap-Nrf-ARE signaling pathway. *Biochem. Biophys. Commun.* **2011**, *410*, 759–765. [CrossRef]
31. Orhan, I.; Özçelik, B.; Kartal, M.; Özdeveci, B.; Duman, H. HPLC quantification of vitexine-2''-O-rhamnoside and hyperoside in three Crataegus species and their antimicrobial and antiviral activities. *Chromatographia* **2007**, *66*, 153–157. [CrossRef]
32. Sun, Y.; Sun, F.; Feng, W.; Qiu, X.; Liu, Y.; Yang, B.; Chen, Y.; Xia, P. Hyperoside inhibits biofilm formation of Pseudomonas aeruginosa. *Exp. Ther. Med.* **2017**, *14*, 1647–1652. [CrossRef]
33. Rodrigo Cavalcante de Araújo, D.; Diego da Silva, T.; Harand, W.; Sampaio de Andrade Lima, C.; Paulo Ferreira Neto, J.; de Azevedo Ramos, B.; Alves Rocha, T.; da Silva Alves, H.; Sobrinho de Sousa, R.; Paula de Oliveira, A.; et al. Bioguided purification of active compounds from leaves of Anadenanthera colubrina var. cebil (Griseb.) Altschul. *Biomolecules* **2019**, *9*, 590. [CrossRef]
34. Fattouch, S.; Caboni, P.; Coroneo, V.; Tuberoso, C.I.; Angioni, A.; Dessi, S.; Marzouki, N.; Cabras, P. Antimicrobial activity of Tunisian quince (*Cydonia oblonga Miller*) pulp and peel polyphenolic extracts. *J. Agric. Food Chem.* **2007**, *55*, 963–969. [CrossRef]
35. Park, S.F. The physiology of Campylobacter species and its relevance to their role as foodborne pathogens. *Int. J. Food Microbiol.* **2002**, *74*, 177–188. [CrossRef]
36. Mingo, E.; Silvan, J.M.; Martinez-Rodriguez, A.J. Selective antibacterial effect on Campylobacter of a winemaking waste extract (WWE) as a source of active phenolic compounds. *LWT Food Sci. Technol.* **2016**, *68*, 418–424. [CrossRef]
37. Liguori, I.; Russo, G.; Curcio, F.; Bulli, G.; Aran, L.; Della-Morte, D.; Gargiulo, G.; Testa, G.; Cacciatore, F.; Bonaduce, D.; et al. Oxidative stress, aging, and diseases. *Clin. Interv. Aging* **2018**, *13*, 757–772. [CrossRef]

38. Reuter, S.; Gupta, S.C.; Chaturvedi, M.M.; Aggarwal, B.B. Oxidative stress, inflammation, and cancer: How are they linked? *Free Radic. Biol. Med.* **2010**, *49*, 1603–1616. [CrossRef]
39. Shin, H.S.; Satsu, H.; Bae, M.J.; Totsuka, M.; Shimizu, M. Catechol groups enable reactive oxygen species scavenging-mediated suppression of PKD-NFkappaB-IL-8 signaling pathway by chlorogenic and caffeic acids in human intestinal cells. *Nutrients* **2018**, *9*, 165. [CrossRef]
40. Bouayed, J.; Rammal, H.; Dicko, A.; Younos, C.; Soulimani, R. Chlorogenic acid, a polyphenol from Prunus domestica (Mirabelle), with coupled anxiolytic and antioxidant effects. *J. Neurol. Sci.* **2007**, *262*, 77–84. [CrossRef]
41. Kim, H.; Hoon Pan, J.; Kim, S.H.; Lee, J.H.; Park, J.W. Chlorogenic acid ameliorates alcohol-induced liver injuries through scavenging reactive oxygen species. *Biochimie* **2018**, *150*, 131–138. [CrossRef]
42. Liang, N.; Dupuis, J.H.; Yada, R.Y.; Kitts, D.D. Chlorogenic acid isomers directly interact with Keap 1-Nrf2 signalling in Caco-2 cells. *Mol. Cell Biochem.* **2019**, *457*, 105–118. [CrossRef]
43. Conforti, F.; Menichini, F. Phenolic compounds from plants as nitric oxide production inhibitors. *Curr. Med. Chem.* **2011**, *18*, 1137–1145. [CrossRef]
44. Bouayed, J.; Bohn, T. Exogenous antioxidants—Double edged swords in cellular redox state: Health beneficial effects at physiologic doses versus deleterious effects at high doses. *Oxid. Med. Cell Longev.* **2010**, *3*, 228–237. [CrossRef]
45. Banerjee, N.; Kim, H.; Talcott, S.T.; Turner, N.D.; Byrne, D.H.; Mertens-Talcott, S.U. Plum polyphenols inhibit colorectal aberrant crypt foci formation in rats: Potential role of the miR-143/protein kinase B/mammalian target of rapamycin axis. *Nutr. Res.* **2016**, *36*, 1105–1113. [CrossRef]

Article

Anti-Biofilm Effect of Selected Essential Oils and Main Components on Mono- and Polymicrobic Bacterial Cultures

Erika Beáta Kerekes [1,*], Anita Vidács [2], Miklós Takó [1], Tamás Petkovits [1], Csaba Vágvölgyi [1], Györgyi Horváth [3], Viktória Lilla Balázs [3] and Judit Krisch [2]

1 Department of Microbiology, Faculty of Science and Informatics, University of Szeged, H-6726 Szeged, Közép fasor 52, Hungary
2 Institute of Food Engineering, Faculty of Engineering, University of Szeged, H-6724 Szeged, Mars tér 7, Hungary
3 Department of Pharmacognosy, University of Pécs, H-7624 Pécs, Rókus utca 2, Hungary
* Correspondence: kerekes@bio.u-szeged.hu

Received: 26 July 2019; Accepted: 10 September 2019; Published: 12 September 2019

Abstract: Biofilms are surface-associated microbial communities resistant to sanitizers and antimicrobials. Various interactions that can contribute to increased resistance occur between the populations in biofilms. These relationships are the focus of a range of studies dealing with biofilm-associated infections and food spoilage. The present study investigated the effects of cinnamon (*Cinnamomum zeylanicum*), marjoram (*Origanum majorana*), and thyme (*Thymus vulgaris*) essential oils (EOs) and their main components, i.e., trans-cinnamaldehyde, terpinen-4-ol, and thymol, respectively, on single- and dual-species biofilms of *Escherichia coli*, *Listeria monocytogenes*, *Pseudomonas putida*, and *Staphylococcus aureus*. In dual-species biofilms, *L. monocytogenes* was paired with each of the other three bacteria. Minimum inhibitory concentration (MIC) values for the individual bacteria ranged between 0.25 and 20 mg/mL, and trans-cinnamaldehyde and cinnamon showed the highest growth inhibitory effect. Single-species biofilms of *L. monocytogenes*, *P. putida*, and *S. aureus* were inhibited by the tested EOs and their components at sub-lethal concentrations. Scanning electron microscopy images showed that the three-dimensional structure of mature biofilms embedded in the exopolysaccharide matrix disappeared or was limited to micro-colonies with a simplified structure. In most dual-species biofilms, to eliminate living cells from the matrix, concentrations exceeding the MIC determined for individual bacteria were required.

Keywords: antibacterial activity; biofilm; polymicrobial biofilm; essential oil; food spoilage

1. Introduction

Surfaces in the food industry provide an excellent substrate for the development of biofilms by spoilage and pathogenic bacteria [1]. These biofilms are typically highly persistent and can increase the likelihood of cross-contamination, potentially leading to food deterioration and/or serious food-borne diseases [2–5].

Biofilms are heterogeneous in vivo, comprising different microorganisms that interact with each other and form a complex multi-species community. Cooperation or competition occurs between the species and it shapes the community by influencing attachment, microcolony formation, and/or resistance to stress conditions [6–10]. Multi-species biofilms may exhibit enhanced fitness and react differently to antimicrobials from monocultures and planktonic cells [11].

Contamination with *Listeria monocytogenes* causes a large number of food poisoning outbreaks annually [12]. According to a study performed by the European Food Safety Authority in 2018, this food-borne pathogen is the leading cause of hospitalization and death in Europe [13]. Listeriosis is frequently associated with fish and fishery products, ready-to-eat salads, and different meat products. Paté, butter, and soft and semi-soft cheeses are also potential carriers of the pathogen [13,14]. Its persistence under different environmental stresses (e.g., low temperature) makes it difficult to eradicate and cross-contamination risk is high [15]. *Staphylococcus aureus*, *Escherichia coli*, and *Pseudomonas putida* strains are also associated with food spoilage and food poisoning outbreaks [16,17]. Moreover, certain food-spoilage bacteria such as *P. putida* can enhance the adhesion, colonization, and biofilm formation of *L. monocytogenes*, whereas others such as *Staphylococcus sciuri* can inhibit it [18,19]. Although the chemical preservatives that are permitted in foods are considered to not cause any side effects, concerns have been raised about the safety of nitrites and sulphites. To date, no conclusive evidence that nitrite is directly carcinogenic has been provided, but at high doses it has been suggested to be a co-carcinogen [20,21].

Against this background, there is increasing demand for healthier products and natural food additives. To meet these demands, researchers have started to examine natural preservatives instead of synthetic ones [22]. Plants are a source of a range of substances with antimicrobial properties, which are promising candidates for the development of new anti-infective agents [23]. Among these, essential oils (EOs) from aromatic and medicinal plants have been a focus of attention in recent decades [24–27]. In addition, the capacity of some EOs to inhibit biofilm formation in mono- and polymicrobial systems has been documented, suggesting their potential utilization as food preservatives and sanitizing agents [28–30]. In this context, targeting polymicrobial cultures with EOs might be effective for reducing the growth and activity of food-related pathogens.

In previous studies, lemon, marjoram, and cinnamon EOs showed inhibitory effects against *E. coli* and *P. putida* biofilms in mixed-culture systems [31,32]. Here the influence of cinnamon, marjoram, and thyme EOs and their major components, namely, trans-cinnamaldehyde, terpinen-4-ol, and thymol, respectively, on the formation of *L. monocytogenes*, *E. coli*, *S. aureus*, and *P. putida* mono- and polymicrobial biofilms was investigated.

2. Materials and Methods

2.1. Bacterial Strains

The bacterial strains used in this study were provided by the Szeged Microbiological Collection (SZMC). The Gram-positive *L. monocytogenes* SZMC 21307 and *S. aureus* SZMC 110007 and the Gram-negative *E. coli* SZMC 0582 and *P. putida* SZMC 291T were used to form mono- and dual-species biofilms.

For pre-culturing and biofilm formation, tryptic soy broth (TSB) containing (in %) peptone from casein 1.7 (Merck; Budapest, Hungary), peptone from soy meal 0.3 (Oxoid; Hampshire, UK), D(+)-glucose 0.25 (VWR; Debrecen, Hungary), NaCl 0.5 (VWR), and K_2HPO_4 0.25 was used. Pre-culturing was performed for 18–20 h at the optimum temperature for the bacteria to achieve high viable cell counts.

Dual-species biofilms were formed in TSB broth with *L. monocytogenes* paired with *E. coli* or *S. aureus* at 37 °C, or with *P. putida* at 30 °C. Quantification of bacteria in the supernatants before and after treatments was done by spreading on selective media (Palcam for *L. monocytogenes*, Chromocult for *E. coli*, Pseudomonas Selective Agar for *P. putida*, and Baird Parker Agar for *S. aureus*). Lab M Limited (Heywood, UK) provided the first three media and Biolab (Budapest, Hungary) provided Baird Parker Agar.

2.2. EOs

EOs of cinnamon (*Cinnamomum zeylanicum*), marjoram (*Origanum majorana*), and thyme (*Thymus vulgaris*) were purchased from Aromax Natural Products Zrt. (Budapest, Hungary). Their main components, i.e., trans-cinnamaldehyde, terpinen-4-ol, and thymol, respectively, were obtained from

Sigma-Aldrich (Munich, Germany). The composition of the oils was determined by GC-MS (Agilent GC: 6850 Series II; MS: 5975C VL MSD; Agilent, Santa Clara, CA, USA) using an Agilent 19091S-433E (Agilent) column at the laboratory of Aromax Natural Products Zrt.

2.3. Determination of Minimum Inhibitory Concentration (MIC)

For the determination of MIC values, the EOs and their components were diluted in liquid culture medium (TSB) in combination with Tween 40 (1%). At the concentration used, Tween 40 had no effect on the viability of the investigated bacteria.

One hundred microliters of 24 h-old cell suspension (10^6 CFU/mL) of each bacterium in liquid culture medium was added to each well of a 96-well microtiter plate, followed by 100 µL of the diluted EO or its major component. Positive controls contained the inoculated growth medium without any EOs or components, and negative controls contained EOs or components in sterile medium. After 24 h of incubation at the appropriate temperature (37 °C for *E. coli*, *L. monocytogenes*, and *S. aureus* and 30 °C for *P. putida*), absorbance was measured at 600 nm (SPECTROstar Nano Spectrophotometer; BMG Labtech, Offenburg, Germany). Absorbance lower than 10% of the positive control samples, i.e., growth inhibition of 90% or more, was considered as the MIC value. Measurements were performed in triplicate.

2.4. Mono- and Dual-Species Biofilm Formation

For biofilm formation, the method described by Peeters et al. [33] was used. Briefly, polystyrene microtiter plates were inoculated with 200 µL of 24-h-old bacterial culture containing cell count of approximately 10^8 CFU/mL. Following 4 h of cell adhesion at the corresponding temperatures, the supernatant was removed, and the plates were rinsed with physiological saline. Subsequently, 200 µL of fresh medium containing the EO or its component to be examined was added at MIC/2 concentration to avoid total growth inhibition. Plates were further incubated for 24 h to allow a biofilm to form. Positive controls contained the inoculated growth medium but without any EOs or components, and negative controls contained EOs or their components in growth medium. Experiments were repeated at least twice, and six parallel measurements were conducted each time.

The inhibition of biofilm formation was detected by the crystal violet staining method. Briefly, after 24 h of treatment, the supernatant was removed, and the wells were rinsed with physiological saline. For fixation of the biofilms, methanol was added, and the supernatant was removed again. Then, 0.1% crystal violet (CV) solution was added to each well and, 20 min later, the excess dye was removed by washing the plates under running tap water. The bound CV was released by adding 200 µL of 33% acetic acid followed by an incubation for 10 min at room temperature. The absorbance was measured at 590 nm (SPECTROstar Nano Spectrophotometer).

For polymicrobial biofilms, except for those produced from the culture containing *L. monocytogenes* and *E. coli*, 24-h-old liquid monocultures containing approximately 10^5 CFU/mL were mixed at a 1:1 ratio in TSB broth. In contrast, the inoculum size for *L. monocytogenes* and *E. coli* was 10^3 CFU/mL because the cell count of the dual culture reached more than 10^{10} CFU/mL after 24 h when a larger inoculum was used. For dual-species biofilms, the EO concentrations were chosen according to the MIC values of the individual strains. For investigation of the possible interactions, for example, synergism or antagonism between the species, concentrations corresponding to half and double the MIC level were examined. When the MIC values were equal for the two species, the levels corresponding to one-quarter of MIC, half of MIC, MIC, and double MIC were used.

2.5. Scanning Electron Microscopy (SEM) Observations

SEM was used to investigate the structural differences between biofilms before and after treatment with selected EOs and their components.

Polymicrobial biofilms were prepared in six-well microtiter plates, where sterile 2 × 2 cm cover slips served as the surfaces to which the cells attached. After 4 h of incubation, the cover slips were rinsed with sterile water, then, fresh medium containing the EOs or their components was added, and the plates were further incubated for 24 h at 37 °C. The concentrations of EOs and components used in this experiment were based on the results on dual-species biofilms.

Preparation of the cover slip-biofilm samples for SEM was performed as described previously [31]. Briefly, samples were immersed in 2.5% glutaraldehyde for 2 h at room temperature and then dehydrated using increasing ethanol concentrations. Final dehydration was performed with *t*-butanol–100% ethanol solution at different ratios, followed by absolute *t*-butanol. After replacing the *t*-butanol with a fresh volume of it and storing the samples at 4 °C for 1 h, they were freeze-dried overnight. Before SEM analysis, a gold membrane was applied, the whole field was examined, and photographs were taken from relevant areas with a Hitachi S4700 scanning electron microscope (Hitachi, Tokyo, Japan). Changes in the three-dimensional structure were mainly visualized with small scale magnification, while higher magnification was used for a more detailed view of the cell wall degrading effect of EOs.

2.6. Statistical Analysis

Data were analyzed by one-way ANOVA followed by Tukey's test or paired *t*-test using GraphPad Prism version 6.0 (GraphPad Software Inc., San Diego, CA, USA. Differences were considered significant at $p < 0.05$.

3. Results

3.1. Composition of EOs and MICs

The compositions of the EOs were previously determined by the producers, and detailed data on them were also reported in previous studies [32,34]. The major components of the EOs were terpinen-4-ol (33.5%) for marjoram, trans-cinnamaldehyde (93.1%) for cinnamon, and thymol (51.8%) for thyme.

MIC values ranged between 0.25 and 20 mg/mL and the bacterium most sensitive to the agents was found to be *E. coli* (Table 1). Among the substances investigated, cinnamon and the components cinnamaldehyde and thymol presented the lowest MIC values. The latter result supports the generally accepted finding that phenolics have the best antimicrobial activity among EO compounds [25,35,36].

Table 1. Minimum inhibitory concentration (MIC) of essential oils (EOs) and their components against the food-related bacteria investigated.

EOs and Components	MIC (mg/mL)			
	Escherichia coli	*Listeria monocytogenes*	*Pseudomonas putida*	*Staphylococcus aureus*
Cinnamon	0.25	1	0.25	0.4
Marjoram	0.5 [1]	4 [2]	2 [1]	3.2
Thyme	1.0	2 [2]	20	0.8
Trans-cinnamaldehyde	0.25	0.25	0.25	0.4
Terpinene-4-ol	2 [1]	4	4 [1]	3.2
Thymol	0.5	0.5 [2]	0.5	0.8

[1] Results from Kerekes et al. (2013). [2] Results from Kerekes et al. (2016).

3.2. Anti-Biofilm-Forming Effect of EOs and Their Major Components

3.2.1. Monocultures

Table 2 summarizes the effects of the tested EOs on monoculture biofilms. For *E. coli*, each components and the thyme EO showed significant inhibitory effect when compared to the control. Cinnamon, despite its notable antibacterial effect, did not reduce the biofilm-forming ability of *E. coli*, but its main component and thyme did. Thymol exhibited the best effect against biofilm formation.

Table 2. Effect of essential oils and components (in MIC/2 concentration) on monoculture biofilm formation of food-related bacteria.

Bacteria	Biofilm Formation (OD_{590}) [1]						
	Positive Control	Cinnamon	Marjoram	Thyme	Trans-cinnamaldehyde	Terpinene-4-ol	Thymol
Escherichia coli	0.68 ± 0.17a	0.77 ± 0.13a	0.72 ± 0.10a [2]	0.34 ± 0.02b	0.32 ± 0.08b	0.44 ± 0.11b [2]	0.14 ± 0.02c
Listeria monocytogenes	1.34 ± 0.14a	0.14 ± 0.02c	0.28 ± 0.04b	0.30 ± 0.04b	0.17 ± 0.02c	0.34 ± 0.34b	0.53 ± 0.08b
Pseudomonas putida	1.52 ± 0.4a	0.48 ± 0.05d	0.08 ± 0.00b [2]	0.51 ± 0.05d	0.70 ± 0.05d	1.09 ± 0.10b [2]	0.43 ± 0.07d
Staphylococcus aureus	2.36 ± 0.07a	1.31 ± 0.30b	1.18 ± 0.20b	1.22 ± 0.16b	1.19 ± 0.17b	1.60 ± 0.30b	1.37 ± 0.41b

[1] Data are presented as mean ± standard deviation of six replicates. Values within a row with different letters are significantly different ($p < 0.05$). [2] Results from Kerekes et al. (2013).

All EOs and their investigated components had considerable anti-biofilm-forming effects on *L. monocytogenes*, as reflected in significant differences compared with the control. Cinnamon and trans-cinnamaldehyde were the best inhibitors and there were no significant differences between the anti-biofilm-forming capacity of the EOs and their main components ($p > 0.05$). Similar results could be seen in the case of *P. putida*, for which cinnamon, thyme, and their major components inhibited biofilm formation, but no significant differences were observed among these groups. Marjoram and terpinene-4-ol also had a strong effect against biofilm formation of all bacteria studied.

For *S. aureus* biofilms, all EOs and components tested proved to be effective against their formation. Absorbance of the treated samples differed significantly from that in the control ($p < 0.001$), but not from each other (Table 2).

3.2.2. Polymicrobial Cultures

L. monocytogenes and *E. coli*

The concentration used for cinnamon was 0.1–2 mg/mL and for trans-cinnamaldehyde 0.06–0.5 mg/mL. After biofilm formation for 24 h, the cell number of control samples was 11 log CFU/mL for *L. monocytogenes* and 5 log CFU/mL for *E. coli*. Treatment with cinnamon at concentrations higher than 0.1 mg/mL decreased the biofilm formation (CFU: 10 log CFU/mL for both bacteria) (Figure 1A). At 0.25 mg/mL (MIC of *E. coli*), both bacteria grew up to 10 log CFU/mL each. At 0.5 mg/mL concentration (double the MIC for *E. coli* and half the MIC for *L. monocytogenes*, see Table 1), a monoculture biofilm appeared with *Listeria* being absent. From a concentration of 1 mg/mL (MIC for *L. monocytogenes*, see Table 1), no survivors of either bacterium were detected. Trans-cinnamaldehyde exhibited similar activity against biofilm formation as compared to cinnamon (Figure 1B).

After treatment with marjoram (concentration range: 0.5–8 mg/mL) and the component terpinen-4-ol (concentration range: 1–8 mg/mL), cell number analysis showed that *E. coli* was eliminated from the biofilm at the lowest concentration, so that this culture contained *L. monocytogenes* cells only ($p < 0.001$). Biofilm elimination started form the lowest concentrations used (0.5 mg/mL for marjoram and 1 mg/mL for terpinene-4-ol) (Figure 1C,D).

Finally, thyme EO (concentration range: 0.5–4 mg/mL) exhibited strong inhibitory effect against the *L. monocytogenes* and *E. coli* mixed culture biofilm, at all the concentrations investigated ($p = 0.03$) (Figure 1E). Moreover, no survivors were detected at 1 mg/mL (half the MIC for *L. monocytogenes*, see Table 1). In addition, thymol (concentration range: 0.1–1 mg/mL) inhibited biofilm formation from its lowest concentration (Figure 1F). Surviving cells were undetectable at the sub-MIC value of 0.2 mg/mL.

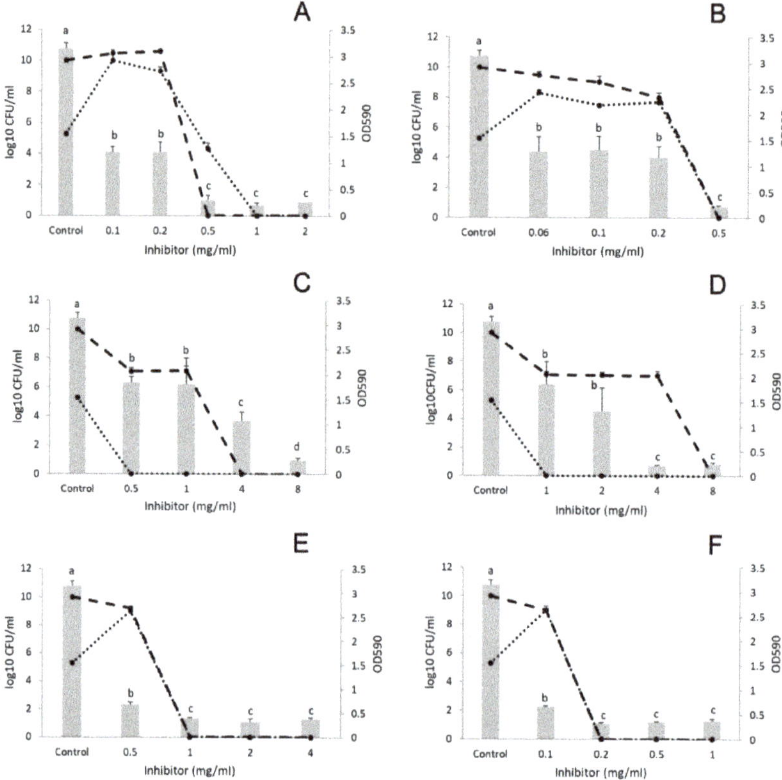

Figure 1. Effect of essential oils (EOs) on the biofilm formation of *Listeria monocytogenes* and *Escherichia coli* polymicrobial cultures. (**A**) cinnamon EO, (**B**) cinnamaldehyde, (**C**) marjoram EO, (**D**) terpinen-4-ol, (**E**) thyme EO, and (**F**) thymol. Columns represent the OD_{590} values, dashed lines represent cell numbers of *L. monocytogenes*, and dotted lines cell numbers of *E. coli*. Results are presented as mean ± standard deviation of six replicates. Different letters indicate statistically significant differences between columns ($p < 0.05$).

L. monocytogenes and *S. aureus*

In the control sample, 7 log CFU/mL was recorded for both bacteria after biofilm formation for 24 h. All EOs and their components significantly decreased biofilm formation (Figure 2). The results indicate that only concentrations higher than the MIC values were effective at eliminating the bacteria. At 8 mg/mL (double the MIC for *Listeria* and more than double the MIC for *S. aureus*, see Table 1), marjoram EO (concentration range: 1.6–8 mg/mL) eliminated *Listeria* from the mixed culture, but *S. aureus* was still present (Figure 2E). At the same concentration, terpinen-4-ol (concentration range: 1.6–8 mg/mL) reduced the CFU of both bacteria to an undetectable level (Figure 2F). Similar results were obtained with cinnamon (concentration range: 0.2–2 mg/mL) and its component trans-cinnamaldehyde (concentration range: 0.1–0.8 mg/mL) (Figure 2A,B). Moreover, thyme (concentration range: 0.4–4 mg/mL) reduced polymicrobial biofilms from a concentration of 0.4 mg/mL (MIC for *L. monocytogenes*) (Figure 2C). Finally, thymol (concentration range: 0.2–1.5 mg/mL) reduced the biofilm investigated and killed both bacteria from a concentration of 0.2 mg/mL (MIC for *Listeria*) (Figure 2D).

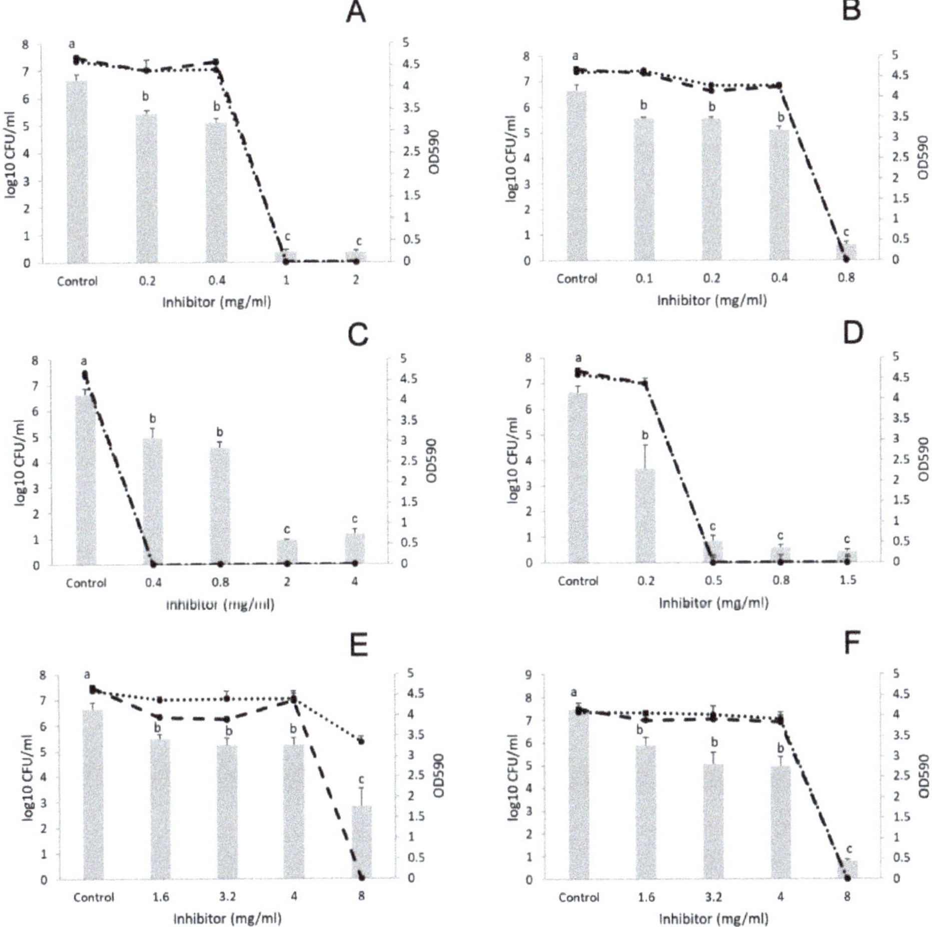

Figure 2. Effect of EOs on the biofilm formation of *Listeria monocytogenes* and *Staphylococcus aureus* polymicrobial cultures. (**A**) cinnamon EO, (**B**) trans-cinnamaldehyde, (**C**) thyme EO, (**D**) thymol, (**E**) marjoram EO, and (**F**) terpinen-4-ol. Columns represent the OD_{590} values, dashed lines represent cell numbers of *L. monocytogenes*, and dotted lines cell numbers of *S. aureus*. Results are presented as mean ± standard deviation of six replicates. Different letters indicate statistically significant differences between columns ($p < 0.05$).

L. monocytogenes and *P. putida*

Figure 3 shows that the number of *Listeria* cells was 8 log CFU/mL and that of *P. putida* was 7 log CFU/mL in the control samples. Concerning cinnamon EO (concentration range: 0.1–1 mg/mL) and trans-cinnamaldehyde (concentration range: 0.1–1 mg/mL), significant reduction in biofilm formation was observed at 0.1 mg/mL; although higher concentrations up to 0.5 mg/mL did not achieve better inhibition (Figure 3A, B). Thyme EO (concentration range 1–20 mg/mL) inhibited the formation of *L. monocytogenes* and *P. putida* co-cultured biofilm as well, at 1 mg/mL (Figure 3C). However, *P. putida* was present (4 log CFU/mL) in the biofilm even at the highest concentration of thyme (20 mg/mL) (Figure 3C), which was equal to the MIC against this bacterium (Table 1). Thymol was applied between 0.1–1 mg/L where biofilm inhibition and decrease in CFU started at the lowest concentration used.

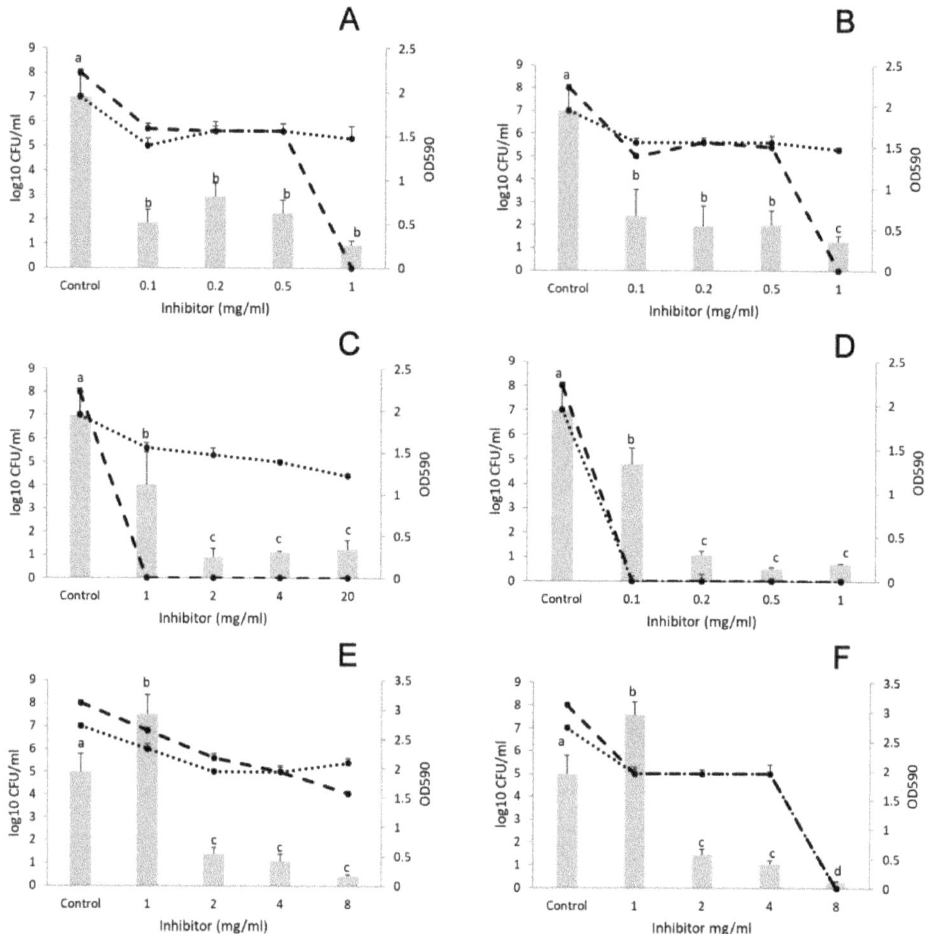

Figure 3. Effect of EOs and EO main components on the biofilm formation of *Listeria monocytogenes* and *Pseudomonas putida* polymicrobial cultures. (**A**) cinnamon EO, (**B**) trans-cinnamaldehyde, (**C**) thyme EO, (**D**) thymol, (**E**) marjoram EO, and (**F**) terpinen-4-ol. Columns represent the OD_{590} values, dashed lines represent cell numbers of *L. monocytogenes*, and dotted lines cell numbers of *P. putida*. Results are presented as mean ± standard deviation of six replicates. Different letters indicate statistically significant differences between columns ($p < 0.05$).

Marjoram EO (concentration range: 1–8 mg/mL) decreased the number of *P. putida* cells at 8 mg/mL by only 2 log CFU/mL compared with the control (double the MIC for *Listeria* and four times the MIC for *P. putida*), (Figure 3E). Meanwhile terpinene-4-ol (concentration range: 1–8 mg/mL) eliminated both bacteria from the biofilm at 8 mg/mL and had similar biofilm inhibitory effect to the parent oil.

3.3. SEM Observations

3.3.1. *L. monocytogenes* and *E. coli*

Cinnamon and trans-cinnamaldehyde were not used in these investigations due to their high inhibitor effect. For *Listeria* and *E. coli* only thyme EO and thymol were chosen based on results obtained for dual-species biofilm.

Control samples showed the complex structure of the formed biofilm (Figure 4A), namely, cells embedded in a large amount of possible extracellular polysaccharide (EPS) and forming a three-dimensional structure. Treatment with thyme and thymol resulted in damaged, anamorph, sparse micro-colonies and individual cells (Figure 4B–D).

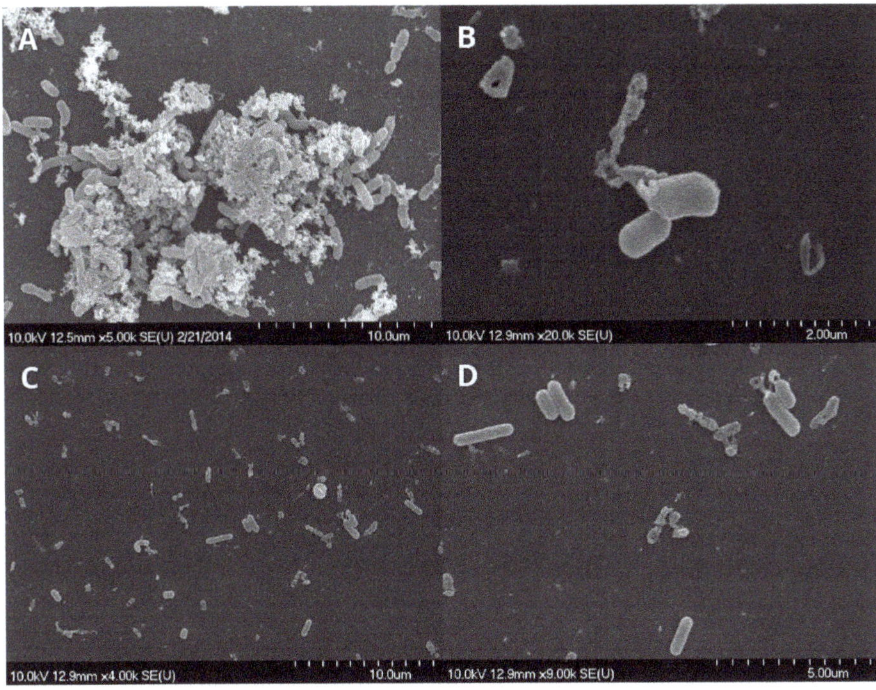

Figure 4. *Listeria monocytogenes* and *Escherichia coli* biofilms observed using scanning electron microscope. (**A**) control, (**B**) thyme EOs (0.5 mg/mL), and (**C**,**D**) thymol (0.1 mg/mL).

3.3.2. *L. monocytogenes* and *S. aureus*

Figure 5A,B present the findings for control samples of *Listeria* and *Staphylococcus* polymicrobial biofilms. The three-dimensional structure embedded in the EPS layer can be observed with *Listeria* cells are at the bottom, to which *Staphylococcus* clusters attach. After treatment with EOs, this structure appears only in micro-colonies with a simple shape, the number of attached cells is decreased, and mainly damaged cells are present. Samples treated with marjoram show split cocci, in agreement with the fact that the EO's target is the cell wall. In some cases, cell debris appears on the surface, which can serve as a possible adhesion site for new cells (Figure 5C–F).

3.3.3. *L. monocytogenes* and *P. putida*

For *Listeria* and *P. putida* only thyme EO and thymol were chosen based on results obtained for dual-species biofilm.

Figure 6A,B present the findings for control samples with nascent and fully formed cell-to-cell connections and EPS material formation. Treatment with marjoram and its major component, terpinen-4-ol, resulted in significant changes in the composition of these structures, leaving mainly damaged cells and cell debris to be observed (Figure 6C,D).

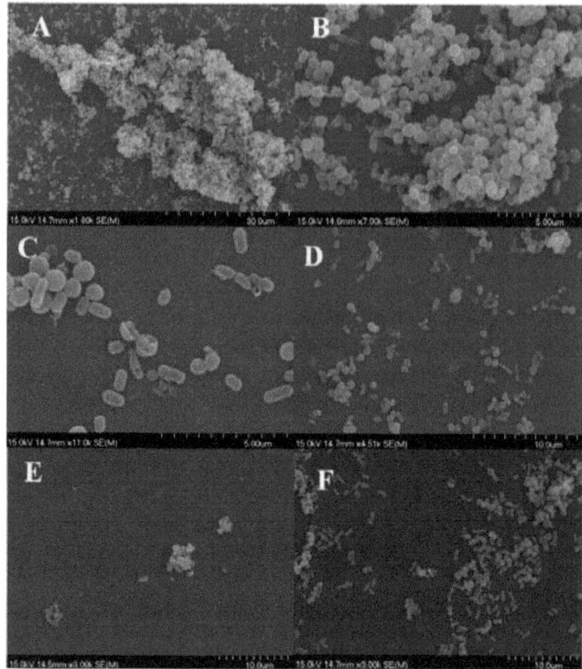

Figure 5. *Listeria monocytogenes* and *Staphylococcus aureus* biofilms observed using scanning electron microscope. (**A**,**B**) control, (**C**) marjoram essential oil (EO) (1.6 mg/mL), (**D**) terpinen-4-ol (8 mg/mL), (**E**) thyme EO (0.4 mg/mL), and (**F**) thymol (0.8 mg/mL).

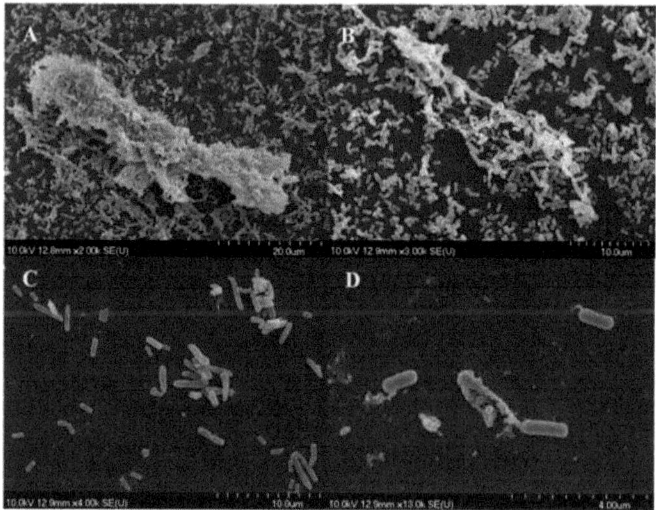

Figure 6. *Listeria monocytogenes* and *Pseudomonas putida* biofilms observed using scanning electron microscope. (**A**,**B**) control, (**C**) marjoram EO (1 mg/mL), and (**D**) terpinen-4-ol (1 mg/mL).

Scanning electron microscopy images demonstrate the structural changes caused by EOs or components during biofilm development [36]. The three-dimensional structure of matured biofilms disappeared after EO treatment, the cells were damaged, and most of the treated cells had burst.

4. Discussion

In the present study we demonstrated that cinnamon, marjoram, and thyme EOs and main components tested had good antibacterial and anti-biofilm forming effect on the investigated bacteria associated with food spoilage and outbreaks. Trans-cinnamaldehyde and thymol were the best inhibitors with MIC values below 1 mg/mL. As expected from the composition pattern of EO cinnamon, the trans-cinnamaldehyde and the EO exhibited similar effect against the bacterial biofilms studied. Besides this, thymol and trans-cinnamaldehyde are phenolics having the best antimicrobial activity among the EO compounds. For comparison, the monoterpene terpiene-4-ol did not exceed the effect of its parent oil marjoram.

Monoculture biofilms were significantly inhibited by the EOs and components in MIC/2 concentration which suggest that growth-reducing effect is not solely responsible for biofilm inhibition. Sub-lethal damage of the cell wall can negatively influence bacterial attachment to surfaces which is the first step in biofilm formation [29].

Dual-species biofilms responded to EOs differently. Contrary to our findings, Almeida et al. [37] reported that, in dual-species (*E. coli*/*L. monocytogenes*) biofilms, cell densities were not altered. In our case, *L. monocytogenes* outgrew *E. coli* in the control samples and most of the EOs eliminated this bacterium from the biofilm. In line with this, the study of Giaouris et al. [9] concluded that *Listeria* formed a stronger biofilm in mixed population, moreover, presence of other bacteria increased its growth. Co-culturing of *L. monocytogenes* and *S. aureus* resulted a strong biofilm, which is in agreement with the results of Millezi et al. [38]; only high concentrations of EOs (e.g., cinnamon EO: 1–2 mg/mL, marjoram EO, and terpinene: 4–8 mg/mL) and components inhibited their formation. *P. putida* proved to be more resistant to the oils and compounds than the *Listeria* and this bacterium was present in the population even at high agent concentrations. This is in accordance with the results of Giaouris et al. [9] who showed that co-culturing with *L. monocytogenes* within a dual-species biofilm increased the community resistance of *P. putida*.

In most cases, anti-biofilm formation effect showed no concentration dependence above a certain concentration, but the results of cell number determination were not congruent with this. In most cases, dual-species biofilms were inhibited at lower concentration than the MIC of the individual bacteria, but cell death occurred mainly at the higher MIC value or above. This discrepancy pointed to the fact that absorbance and cell enumeration data can differ significantly. Biofilms, inhibited to a high degree based on absorbance data, can still contain enough living cells to cause hygiene problems. Scanning electron microscopic images demonstrated the structural changes caused by EOs or components during biofilm development. Similarly to the altered biofilm structure visualized by confocal laser scanning electron microscope in a recent study [39], our SEM pictures also showed that the three-dimensional structure of matured biofilms disappeared after EO treatment and most of the treated cells have been burst. Studying the mechanism of EOs in more detail, Zhang et al. [40] detected leakage of electrolytes due to disruption of cell permeability after EO treatment which eventually lead to cell death. Along with our findings these results also support the fact that EOs induce severe membrane damage [25,40].

In conclusion, the EOs and EO components examined in this study could represent alternatives for elimination of *E. coli*, *L. monocytogenes*, *P. putida*, and *S. aureus* single and *L. monocytogenes*-*E. coli*/*S. aureus*/*P. putida* dual biofilms in vitro with different efficiency. The use of EOs as antimicrobial agents in real food systems is often limited due to their strong odor and taste. Therefore, our future investigations aim at novel approaches, such as encapsulation of EOs, that could potentially reduce the organoleptic impact and increase the antimicrobial activity [41].

Author Contributions: Conceptualization, E.B.K, J.K., and C.V.; methodology and data analysis, E.B.K., A.V., M.T., T.P., G.H., and V.L.B.; writing—original draft preparation, E.B.K.; writing—review and editing, J.K., C.V., and M.T.

Funding: This work was supported by the Hungarian Government and the European Union within the framework of the Széchenyi 2020 Programme through grants GINOP-2.3.2-15-2016-00012, GINOP-2.3.3-15-2016-00006 and EFOP-3.6.1-16-2016-00008. The research of M.T. was supported by grants through the János Bolyai Research Scholarship of the Hungarian Academy of Sciences and the UNKP-18-4 New National Excellence Program of the Ministry of Human Capacities. The research of G.H. was supported by the NKFI K 128217 Research Scholarship.

Conflicts of Interest: The authors declare no conflict of interest. The funders had no role in the design of the study; in the collection, analyses, or interpretation of data; in the writing of the manuscript, or in the decision to publish the results.

References

1. Shi, X.; Zhu, X. Biofilm formation and food safety in food industries. *Trends Food Sci. Technol.* **2009**, *20*, 407–413. [CrossRef]
2. Midelet, G.; Carpentier, B. Transfer of microorganisms, including *Listeria monocytogenes*, from various materials to beef. *Appl. Environ. Microbiol.* **2002**, *68*, 4015–4024. [CrossRef]
3. Johnson, L.R. Microcolony and biofilm formation as a survival strategy for bacteria. *J. Theor. Biol.* **2008**, *251*, 24–34. [CrossRef]
4. Wong, H.S.; Townsend, K.M.; Fenwick, S.G.; Trengove, R.D.; O'Handley, R.M. Comparative susceptibility of planktonic and 3-day-old *Salmonella Typhimurium* biofilms to disinfectants. *J. Appl. Microbiol.* **2010**, *108*, 2222–2228. [CrossRef]
5. Bridier, A.; Briandet, R.; Thomas, V.; Dubois-Brissonnet, F. Resistance of bacterial biofilms to disinfectants: A review. *Biofouling* **2011**, *27*, 1017–1032. [CrossRef]
6. Hall-Stoodley, L.; Costerton, J.W.; Stoodley, P. Bacterial biofilms: From the natural environment to infectious diseases. *Nat. Rev. Microbiol.* **2004**, *2*, 95–108. [CrossRef]
7. Rendueles, O.; Ghigo, J.M. Multi-species biofilms: How to avoid unfriendly neighbours. *FEMS Microbiol. Rev.* **2012**, *36*, 972–989. [CrossRef]
8. Ica, T.; Caner, V.; Istanbullu, O.; Nguyen, H.D.; Ahmed, B.; Call, D.R.; Beyenal, H. Characterization of mono- and mixed-culture *Campylobacter jejuni* biofilm. *Appl. Environ. Microbiol.* **2012**, *78*, 1033–1038. [CrossRef]
9. Giaouris, E.; Chorianopoulos, N.; Doulgeraki, A.; Nychas, G.J. Co-culture with *Listeria monocytogenes* within a dual-species biofilm community strongly increases resistance of *Pseudomonas putida* to benzalkonium chloride. *PLoS ONE* **2013**, *8*, e77276. [CrossRef]
10. Jahid, I.K.; Han, N.R.; Srey, S.; Ha, S.-D. Competitive interactions inside mixed-culture biofilms of *Salmonella Typhimurium* and cultivable indigenous microorganisms on lettuce enhance microbial resistance of their sessile cells to ultraviolet C (UV-C) irradiation. *Food Res. Int.* **2014**, *55*, 445–454. [CrossRef]
11. Elias, S.; Banin, E. Multi-species biofilms: Living with friendly neighbors. *FEMS Microbiol. Rev.* **2012**, *36*, 990–1004. [CrossRef]
12. Thakur, M.; Asrani, R.K.; Patial, V. Listeria monocytogenes: A Food-Borne Pathogen. In *Foodborne Diseases. Handbook of Food Bioengineering*; Holban, A.A., Grumezescu, A.M., Eds.; Academic Press: London, UK, 2018; Volume 15, pp. 157–192. [CrossRef]
13. European Food Safety Authority. The European Union summary report on trends and sources of zoonoses, zoonotic agents and food-borne outbreaks in 2017. *EFSA J.* **2018**, *16*, 5500. [CrossRef]
14. Liu, D. Identification, subtyping and virulence determination of *Listeria monocytogenes*, an important foodborne pathogen. *J. Med. Microbiol.* **2006**, *55*, 645–659. [CrossRef]
15. Puga, C.H.; SanJose, C.; Orgaz, B. Biofilm development at low temperatures enhances Listeria monocytogenes resistance to chitosan. *Food Contr.* **2016**, *65*, 143–151. [CrossRef]
16. Fetsch, A.; Contzen, M.; Hartelt, K.; Kleiser, A.; Maassen, S.; Rau, J.; Kraushaar, B.; Layer, F.; Strommenger, B. *Staphylococcus aureus* food-poisoning outbreak associated with the consumption of ice-cream. *Int. J. Food Microbiol.* **2014**, *187*, 1–6. [CrossRef]
17. Naimi, T.; Wicklund, J.; Olsen, S.J.; Krause, G.; Wells, J.G.; Bartkus, J.M.; Boxrud, D.J.; Sullivan, M.; Kassenborg, H.; Besser, J.M.; et al. Concurrent outbreaks of *Shigella sonnei* and enterotoxigenic *Escherichia coli* infections associated with parsley: Implications for surveillance and control of foodborne illness. *J. Food Prot.* **2003**, *66*, 535–541. [CrossRef]

18. Leriche, V.; Carpentier, B. Limitation of adhesion and growth of *Listeria monocytogenes* on stainless steel surfaces by *Staphylococcus sciuri* biofilms. *J. Appl. Microbiol.* **2000**, *88*, 594–605. [CrossRef]
19. Carpentier, B.; Chassaing, D. Interactions in biofilms between *Listeria monocytogenes* and resident microorganisms from food industry premises. *Int. J. Food Microbiol.* **2004**, *97*, 111–122. [CrossRef]
20. Schweinsberg, F.; Bürkle, V. Nitrite: A co-carcinogen? *J. Cancer Res. Clin. Oncol.* **1985**, *109*, 200–202. [CrossRef]
21. Cammack, R.; Joannou, C.L.; Cui, X.Y.; Torres Martinez, C.; Maraj, S.R.; Hughes, M.N. Nitrite and nitrosyl compounds in food preservation. *BBA-Bioenergetics* **1999**, *1411*, 475–488. [CrossRef]
22. Mariutti, L.R.B.; Nogueira, G.C.; Bragagnolo, N. Lipid and cholesterol oxidation in chicken meat are inhibited by sage but not by garlic. *J. Food Sci.* **2011**, *76*, C909–C915. [CrossRef]
23. Schelz, Z.; Hohmann, J.; Molnar, J. Recent advances in research of antimicrobial effects of essential oils and plant derived compounds on bacteria. In *Ethnomedicine: A Source of Complementary Therapeutics*; Chattopadhyay, D., Ed.; Research Signpost: Kerala, India, 2010; pp. 179–201.
24. Burt, S. Essential oils: Their antibacterial properties and potential applications in foods—A review. *Int. J. Food Microbiol.* **2004**, *94*, 223–253. [CrossRef]
25. Bakkali, F.; Averbeck, S.; Averbeck, D.; Idaomar, M. Biological effects of essential oils–a review. *Food Chem. Toxicol.* **2008**, *46*, 446–475. [CrossRef]
26. Tajkarimi, M.M.; Ibrahim, S.A.; Cliver, D.O. Antimicrobial herb and spice compounds in food. *Food Contr.* **2010**, *21*, 1199–1218. [CrossRef]
27. Sánchez, E.; García, S.; Heredia, N. Extracts of edible and medicinal plants damage membranes of *Vibrio cholerae*. *Appl. Environ. Microbiol.* **2010**, *76*, 6888–6894. [CrossRef]
28. Chorianopoulos, N.G.; Giaouris, E.D.; Skandamis, P.N.; Haroutounian, S.A.; Nychas, G.J.E. Disinfectant test against monoculture and mixed-culture biofilms composed of technological, spoilage and pathogenic bacteria: Bactericidal effect of essential oil and hydrosol of *Satureja thymbra* and comparison with standard acid-base sanitizers. *J. Appl. Microbiol.* **2008**, *104*, 1586–1596. [CrossRef]
29. Oliveira, M.M.M.; Brugnera, D.F.; Cardoso, M.G.; Alves, E.; Piccoli, R.H. Disinfectant action of *Cymbopogon sp.* essential oils in different phases of biofilm formation by *Listeria monocytogenes* on stainless steel surface. *Food Contr.* **2010**, *21*, 549–553. [CrossRef]
30. Adukwu, E.C.; Allen, S.C.H.; Phillips, C.A. The anti-biofilm activity of lemongrass (*Cymbopogon flexuosus*) and grapefruit (*Citrus paradisi*) essential oils against five strains of *Staphylococcus aureus*. *J. Appl. Microbiol.* **2012**, *113*, 1217–1227. [CrossRef]
31. Kerekes, E.B.; Deák, É.; Takó, M.; Tserennadmid, R.; Petkovits, T.; Vágvölgyi, C.; Krisch, J. Anti-biofilm forming and anti-quorum sensing activity of selected essential oils and their main components on food-related micro-organisms. *J. Appl. Microbiol.* **2013**, *115*, 933–942. [CrossRef]
32. Kerekes, E.B.; Vidács, A.; Jenei Török, J.; Gömöri, C.; Petkovits, T.; Chandrashekaran, M.; Kadaikunnan, S.; Alharbi, N.S.; Vágvölgyi, C.; Krisch, J. Anti-listerial effect of selected essential oils and thymol. *Acta Biol. Hung.* **2016**, *67*, 333–343. [CrossRef]
33. Peeters, E.; Nelis, H.J.; Coenye, T. Comparison of multiple methods for quantification of microbial biofilms grown in microtiter plates. *J. Microbiol. Meth.* **2008**, *72*, 157–165. [CrossRef]
34. Gömöri, C.; Vidács, A.; Kerekes, E.B.; Nacsa-Farkas, E.; Böszörményi, A.; Vágvölgyi, C.; Krisch, J. Altered antimicrobial and anti-biofilm forming effect of thyme essential oil due to changes in composition. *Nat. Prod. Commun.* **2018**, *13*, 483–487. [CrossRef]
35. Dorman, H.J.D.; Deans, S.G. Antimicrobial agents from plants: Antibacterial activity of plant volatile oils. *J. Appl. Microbiol.* **2000**, *88*, 308–316. [CrossRef]
36. Nazzaro, F.; Fratianni, F.; De Martino, L.; Coppola, R.; De Feo, V. Effect of essential oils on pathogenic bacteria. *Pharmaceuticals* **2013**, *6*, 1451–1474. [CrossRef]
37. Almeida, C.; Azevedo, N.F.; Santos, S.; Keevil, C.W.; Vieira, M.J. Discriminating multi-species populations in biofilms with peptide nucleic acid fluorescence in situ hybridization (PNA FISH). *PLoS ONE* **2011**, *6*, e14786. [CrossRef]
38. Millezi, F.M.; Pereira, M.O.; Batista, N.N.; Camargos, N.; Auad, I.; Cardoso, M.D.G.; Piccoli, R.H. Susceptibility of monospecies and dual-species biofilms of *Staphylococcus aureus* and *Escherichia coli* to essential oils. *J. Food Saf.* **2012**, *32*, 351–359. [CrossRef]

39. Szczepanski, S.; Lipski, A. Essential oils show specific inhibiting effects on bacterial biofilm formation. *Food Contr.* **2014**, *36*, 224–229. [CrossRef]
40. Zhang, Y.; Liu, X.; Wang, Y.; Jiang, P.; Quek, S. Antibacterial activity and mechanism of cinnamon essential oil against *Escherichia coli* and *Staphylococcus aureus*. *Food Contr.* **2016**, *59*, 282–289. [CrossRef]
41. Hyldgaard, M.; Mygind, T.; Meyer, R.L. Essential oils in food preservation: Mode of action, synergies, and interactions with food matrix components. *Front. Microbiol.* **2012**, *3*, 12. [CrossRef]

Article

Evaluation of the Potential of Biofilm Formation of *Bifidobacterium longum* subsp. *infantis* and *Lactobacillus reuteri* as Competitive Biocontrol Agents Against Pathogenic and Food Spoilage Bacteria

Barbara Speranza [1], Arcangelo Liso [2,*], Vincenzo Russo [3] and Maria Rosaria Corbo [1,*]

1 Department of the Science of Agriculture, Food and Environment (SAFE), University of Foggia, Via Napoli 25, 71122 Foggia, Italy; barbara.speranza@unifg.it
2 Department of Medicine and Surgery, University of Foggia, Polo Biomedico, Viale Pinto 1, 71122 Foggia, Italy
3 Institute of Ophthalmology, Department of Surgery Science, University of Foggia, Viale Pinto, 71122 Foggia, Italy; virus66@alice.it
* Correspondence: arcangelo.liso@unifg.it (A.L.); mariarosaria.corbo@unifg.it (M.R.C.)

Received: 29 November 2019; Accepted: 21 January 2020; Published: 25 January 2020

Abstract: This study proposes to exploit the in vivo metabolism of two probiotics (*Bifidobacterium longum* subsp. *infantis* and *Lactobacillus reuteri*) which, upon adhesion on a solid surface, form a biofilm able to control the growth of pathogenic and food spoilage bacteria. The results showed that pathogenic cell loads were always lower in presence of biofilm (6.5–7 log CFU/cm^2) compared to those observed in its absence. For *Escherichia coli* O157:H7, a significant decrease (>1–2 logarithmic cycles) was recorded; for *Listeria monocytogenes*, *Staphylococcus aureus*, and *Salmonella enterica*, cell load reductions ranged from 0.5 to 1.5 logarithmic cycles. When tested as active packaging, the biofilm was successfully formed on polypropylene, polyvinyl chloride, greaseproof paper, polyethylene and ceramic; the sessile cellular load ranged from 5.77 log CFU/cm^2 (grease-proof paper) to 6.94 log CFU/cm^2 (polyethylene, PE). To test the potential for controlling the growth of spoilage microorganisms in food, soft cheeses were produced, inoculated with *L. monocytogenes* and *Pseudomonas fluorescens*, wrapped in PE pellicles with pre-formed biofim, packed both in air and under vacuum, and stored at 4 and 15 °C: an effective effect of biofilms in slowing the decay of the microbiological quality was recorded.

Keywords: probiotic; biofilm; pathogen; spoilage bacteria; active packaging

---

1. Introduction

Despite that their use in foods is dated, in the last decades, Lactic Acid Bacteria (LAB) have attracted much attention for their documented beneficial properties and for potential useful applications. Among LAB, several strains are currently claimed as probiotics [1], i.e., live microorganisms that, when administered in adequate amounts, confer a health benefit on the host [2]. According to the consensus statement, there are some bacterial species with a long history of safe use and a well-recognized health effect, such as *Bifidobacterium adolescentis*, *B. animalis*, *B. bifidum*, *B. breve*, *B. longum*, *Lactobacillus acidophilus*, *L. reuteri*, *L. casei*, *L. fermentum*, *L. gasseri*, etc. [2]; some strains, such as *B. longum* subsp. *infantis* and *L. reuteri*, are widespread due to the strong evidence of their effect on health [3]. Probiotics are able to colonize, stably or transiently, host mucosal surfaces, including the gut, where they may contribute to host health; the capacity of probiotics to colonize biotic and abiotic surfaces by forming structured communities (i.e., biofilms), could have great potentials for human health and food safety biotechnologies, although this aspect has is in fact barely been explored. It has recently been shown that microbial biofilms may play several "useful" roles such as biodegradation of toxic compounds and pollutants, bioremediation, toxic effluents treatment [4], despite being initially considered only a negative phenomenon. These applications suggest that microbial biofilms could be successfully used for new applications in the biomedical, industrial, food, and environmental field [4].

In the biomedical field, for example, a biofilm formed by probiotic microorganisms could be potentially useful to hinder the development of microorganisms responsible for infections, especially those caused by microorganisms of hospitals, typically resistant to common antibiotic treatments. Indeed, it is widely accepted that in the development of direct and airborne transmission of nosocomial infections, the hospital environment (infection reservoir) plays a key role [5]. In fact, it can be anticipated that a probiotic biofilm left to form ad hoc on several surfaces (e.g., toilets, air conditioning systems) could reduce the spread of pathogenic species that may harbor thereon. Other potential applications in the biomedical field could be: preparations used in skin lesions for the healing processes to add antibacterial capacity, the coating of implants and catheters, medical devices applied to the oral cavity which might hinder the growth of bacterial species associated with caries and periodontal disease [6–9].

On the other hand, regarding potential applications in the food industry, biofilms can be used to ensure the hygienic-sanitary safety of food products, as well as an extension of their shelf-life. The formation of biofilms by "useful/probiotic" microorganisms may be stimulated on materials commonly used to package food (plastic films, pellicles, combinations for packaging, paper, etc.) in order to develop an innovative active packaging system. Although the scientific community is very active in the production of research related to the ability of microorganisms to form biofilms, most studies have focused on biofilm formation by pathogens and/or spoilage microorganisms (*Enterobacter*, *Listeria*, *Micrococcus*, *Streptococcus*, *Bacillus* and *Pseudomonas*) [10–13]. It has been also shown that certain species of LAB are able to form biofilms and some of them are capable of exhibiting antimicrobial activity against pathogenic microorganisms [14–16]; some research was conducted on the possibility of using new methods of sanitation, exploiting the principle of biological competition using probiotic products [17], but this aspect needs to be explored further. In a previous study, we have described the optimization of the production of a probiotic biofilm through intermediate steps by fixing some valuable key points about the probiotics' ability to adhere to surfaces and to form biofilms [18]. These results were used to file a patent covering the use of probiotic biofilms as a means to control pathogen growth [19]. Even if some studies in literature present the use of LAB (mainly lactobacilli) biofilms to control pathogen growth in food and superficies [20–27], most of them propose the use of bio-surfactants and compounds with antimicrobial activity produced in greater quantities by lactobacilli when growing in sessile form. Indeed, our study proposes a probiotic biofilm that exploits the in vivo metabolism of two selected probiotic strains able to adhere rapidly on abiotic surfaces, and not the substances secreted by them and subsequently recovered and used, as in the prior art. To the best of our knowledge, only one study has previously proposed a similar approach evaluating the use of potential probiotic LAB (isolated from Brazilian's foods) biofilms to control *Listeria monocytogenes*, *Salmonella* Typhimurium, and *Escherichia coli* O157:H7 biofilms formation and suggesting that LAB strains can be excellent candidates to form protective biofilms to be used as biocontroller of contamination into the food chain [28].

Besides the use as an innovative active packaging to ensure the safety of food products, as well as an extension of their shelf-life, the proposed probiotic biofilm formed ad hoc on medical devices (catheters, implants, braces, bite blocks or condoms) and on bathrooms' surfaces (sink, bidet, toilet bowl, water closet or piece of furniture) could be considered a tool against colonizing strains, since these surfaces are often implicated in nosocomial infections. Our proposal could lead to the development of a useful means to control the growth of pathogenic and spoilage bacteria for industrial and medical applications. In the following, some specific applications of the developed probiotic biofilm are described, focusing on two different aspects: 1) effect of probiotic biofilms on pathogen sessile growth; 2) application as potential active packaging.

2. Materials and Methods

2.1. Effect of Probiotic Biofilms on Pathogen Sessile Growth

2.1.1. Surfaces and Microorganisms

Polycarbonate resin (Lexan, Fedele s.r.l., Rome, Italy) was the surface chosen for the adhesion experiments. Before each experiment, the chips (2.5 × 5.0 × 0.05 cm) were prepared by washing in acetone for a minimum of 30 min, rinsing in distilled water, and then soaking in 1 N NaOH for 1 h. After a final rinse in distilled water, the chips were allowed to air dry. This cleansing procedure was required to remove fingerprints, oils grease, and other soils that may have been on the materials. The cleaned chips were finally autoclaved at 121 °C for 15 min prior to use.

The probiotic strains used for this study were *Bifidobacterium longum* subsp. *infantis* DSM20088 and *Lactobacillus reuteri* DSM20016, both purchased from Leibniz-Institut DSMZ (Deutsche Sammlung von Mikroorganismen und Zellkulturen) and stored at −20 °C in MRS broth (Oxoid, Milan, Italy).

Before each assay, they were grown in their optimal media at their optimal conditions, until late exponential phase was attained; namely, MRS broth added with cysteine 0.05% (w/v) (Sigma-Aldrich, Milan, Italy) incubated at 37 °C for 24 to 48 h, under anaerobic conditions, and MRS broth (Oxoid) incubated at 30 °C for 24 to 48 h, under anaerobic conditions, were used for *B. infantis* DSM20088 and *L. reuteri* DSM20016, respectively.

Cells cultures were successively harvested by centrifugation for 10 min at 4500 rpm (4 °C) and the pellets were washed twice with sterile saline solution (0.9% NaCl) at 4 °C and finally resuspended in the same solution at a cell concentration of 1×10^8 CFU/mL.

As pathogen targets were chosen, four strains belonging to the Culture Collection of the Laboratory of Predictive Microbiology (Department of the Science of Agriculture, Food and Environment, Foggia University), and microorganisms with the media and growth conditions used, have been listed in Table 1. The organisms were transferred to fresh Nutrient Agar (NA, Oxoid) periodically to maintain viability and, prior to use, they were activated by two successive 24-h transfers of cells in Nutrient broth (NB, Oxoid) at 37 °C. Inocula for experiments were prepared by centrifugation of the 24-h microbial cultures at 3000× *g* for 15 min at 4 °C. After centrifugation, the obtained pellets were resuspended in sterile saline solution at 4 °C to obtain approximately 10^8 CFU/mL for each microorganism.

Table 1. Pathogen strains used in the study with the indication of their source and optimal media and growth conditions adopted.

Strains	Source	Optimal Media and Growth Conditions
Listeria monocytogenes *	Culture Collection of the Laboratory of Predictive Microbiology, SAFE, University of Foggia	Listeria selective agar base (Oxoid) plus Listeria selective supplement-Oxoid formulation, incubated at 37 °C for 48 h
Escherichia coli O157:H7	CECT 4267	Sorbitol MacConkey Agar (Oxoid), incubated at 37 °C for 24 h
Staphylococcus aureus	ATCC 25923	Baird-Parker Agar Base (Oxoid) plus Egg Yolk Tellurite Emulsion, incubated at 37 °C for 24 h
Salmonella enterica	ATCC 35664	Chromatic Salmonella Agar (Liofilchem, Roseto degli Abruzzi, Teramo, Italy), incubated at 37 °C for 24 h

* The strain was isolated from fish products and identified by sequencing the 16SrDNA.

2.1.2. Experiment

Biofilm formation was favoured by simultaneously inoculating the cocktail of identified probiotics (*B. infantis* DSM20088 and *L. reuteri* DSM20016, about ~10^8 CFU/mL) and the pathogenic target (~10^7 CFU/mL) on polycarbonate surfaces (Lexan® tiles, 25 mm × 75 mm, 0.5 mm thick) left at room temperature (20 °C) for 2 h. After this time interval, the tiles were transferred to aliquots of peptone water (1% bacteriological peptone) and incubated at 15 °C for 48 h [18,19]. Specifically, for each pathogen, two samples were prepared: an ACTIVE sample (ACT), containing a chip where probiotics were left to form biofilm; a CONTROL sample (CNT), containing a chip without probiotics. The pathogen sessile cell load was determined after 0, 4, 24, 30 and 48 h after inoculation. At these times, chips were aseptically removed and rinsed with sterile distilled water, in order to eliminate the unattached cells. As suggested in literature [29], sessile cells were detached from chips in a sterile test tube containing 45 mL of sterile saline with a 20 Hz "Vibra Cell" sonicator (SONICS, Newcastle, Conn., USA) for 3 min. Viable and cultivable cells were enumerated by serial dilutions in 0.9% NaCl solution and plating on appropriate media (Table 1).

2.2. Application as Potential Active Packaging

2.2.1. Probiotic Biofilm Formation on Different Materials

The materials assayed were polypropylene (PP), polyvinyl chloride (PVC), greaseproof paper (GP), waxed paper (WP), polyethylene (PE) and ceramic; all materials were cut in rectangles of 2.5 × 5.0 cm and cleaned by immersion in ethanol. Each individual chip was well rinsed with ultrapure water and dried at room temperature. Probiotic biofilms were left to form for 96 h by simultaneously inoculating the cocktail of probiotics (*B. infantis* DSM20088 and *L. reuteri* DSM20016, about ~10^8 CFU/mL) on chips of different materials, left at room temperature (20 °C) for 2 h. After this time interval, the chips were transferred to aliquots of peptone water (1% bacteriological peptone) and incubated at 15 °C for 96 h [19].

Biofilm cells were enumerated at 2, 24, and 96 h after inoculation. At these times, chips were aseptically removed and rinsed with sterile distilled water, in order to eliminate the unattached cells. Sessile cells were detached from chips in a sterile test tube containing 45 mL of sterile saline with a "Vibra Cell" sonicator (SONICS, Newcastle, Conn., USA) at 20 kHz for 3 min. Viable and cultivable cells were enumerated by serial dilutions in 0.9% NaCl solution and plating on MRS Agar (Oxoid).

2.2.2. Challenge Tests

Inoculations for experiments were prepared by centrifugation of the 24-h microbial cultures in an ALC 4239R centrifuge (ALC, Milan, Italy) at 3000× *g* for 15 min at 4 °C. For the inoculations of challenge tests, after centrifugation the pellets were resuspended in sterile isotonic solution (0.9% NaCl) at a temperature of 4 °C and serial dilutions were made to obtain approximately 10^4 CFU/mL for each microorganism. For biofilm formation, the probiotic pellet was resuspended in sterile isotonic solution at a temperature of 4 °C and used on polyethylene films to form experimental pellicles with pre-formed probiotic biofilm (EXP).

Miniature soft cheeses were made using pasteurized, whole and homogenized milk, purchased in a local market. The milk had the following characteristics: lactose 5.0%, protein 3.2%, fat 3.6%, pH 6.6. The cheeses were produced using a domestic cheese-maker ("Casaro", Philips, Milan, Italy) by pouring the milk into the single-wall cheese-maker vessel and heating to 85 °C. As soon as the temperature reached 85 °C (after a few minutes), 4 g/L of sodium chloride was added and the salted milk was immediately left to cool to 30 °C. Renneting was performed with 3 mL/L of liquid calf rennet (concentrate extract of Liquid Rennet, CHR. Hansen s.p.a., Milan, Italy). After coagulation and curd strengthening (approximately 40 min), the curd was cut and the whey discarded. Finally, miniature soft cheeses of a round shape (25 g, 6 cm diameter) and regular smooth surfaces were made by hand and placed in sterile boxes fitted with a grid to facilitate whey draining. The boxes were kept at room temperature for 6 h until packaging.

To test the potential for probiotic biofilms to control the growth of microorganisms in soft cheese, they were inoculated with *L. monocytogenes* (challenge test A) and *Ps. fluorescens* (challenge test B). The inoculation (about 10^2 CFU/g) was carried out in the most homogeneous way possible, spreading 0.2 mL of the prepared microbial suspension across the entire surface of miniature cheese by means of a sterile spatula. After inoculation, all cheeses were wrapped in polyethylene films (EXP) and packed in high-barrier plastic bags (Nylon/Polyethylene, 102 μm (Tecnovac, San Paolo D'Argon, Bergamo, Italy)) by means of S100-Tecnovac equipment. Control batches were prepared by wrapping cheeses in pellicles without pre-formed biofilm (CNT). All samples were packaged in air and under vacuum. During the storage at 4 and 15 °C for 28 and 14 days, respectively, microbiological analyses, determination of pH and measurements of a_w were made, details of which are given below.

2.2.3. Microbiological, Chemico-Physical and Sensorial Analyses

For microbiological analyses, mini-cheeses (25 g) were diluted with 225 mL of 0.1% peptone water with salt (0.9% NaCl) in a Stomacher bag (Seward, London, England) and homogenized for 1 min in a Stomacher Lab Blender 400 (Seward). Serial dilutions of cheese homogenates were plated on the surface of the appropriate media in Petri dishes. The media and the conditions used were: *Listeria* selective agar base (Oxoid) plus *Listeria* selective supplement-Oxoid formulation, incubated at 37 °C for 48 h, for *L. monocytogenes*; *Pseudomonas* Agar Base (Oxoid) plus *Pseudomonas* CFC Supplement, incubated at 25 °C for 48 h, for *Ps. fluorescens*; MRS agar (Oxoid), incubated at 30 °C for 4 days under anaerobiosis, for mesophilic lactobacilli.

For each batch, the measurement of pH was performed twice on the first homogenized dilution of the cheese samples during storage with a Crison pH meter model micro pH 2001 (Crison). a_w was measured by a hygrometer AQUALAB CX-2 (Decagon Device, Pullman, WA, USA).

During the storage at 4 and 15 °C, a sensory evaluation was also performed: the panel consisted of 15 panelists aged between 22 and 38 years (students and researchers of the Department of the Science of Agriculture, Food and Environment (SAFE), University of Foggia). Using a scale ranging from 0 to 10 (where 10 stands for the most attractive attributes and 0 for the absolutely unpleasant attributes), the sensorial overall quality of the samples was determined by evaluating colour, odour, texture and overall acceptability. During the test sessions, cheese samples were coded by a letter and presented individually to each panelist in plastic cups covered with a lid in random order. Sensory evaluation was conducted in individual booths under controlled conditions of light (white light), temperature (20 ± 2 °C), and humidity (70% to 85%).

2.2.4. Statistical Analyses

All experiments were performed twice with the analyses conducted twice.

Results about the effect of probiotic biofilms on pathogen sessile growth were expressed as log CFU/cm^2, presented as the average of replicates (n = 4) and analyzed through the Student's t-test ($p < 0.05$).

To highlight the effectiveness of probiotic biofilm, for each time of analysis the pathogen sessile data were expressed as follows:

$$\text{Biofilm Efficacy} = \text{CNT} - \text{ACT}$$

where CNT and ACT were pathogen cell numbers (log CFU/cm^2) in the control (without probiotic biofilm) and in the active sample (with probiotic biofilm). These differences were analysed through one-way ANOVA and Tukey's test as the post-hoc comparison test ($p < 0.05$). Results about biofilm formation on different materials were expressed as log CFU/cm^2, presented as the average of replicates (n = 4) and analyzed through one-way ANOVA and Tukey's test as the post-hoc comparison test ($p < 0.05$).

The microbiological data collected during the challenge tests were expressed as the average of two replicates and the obtained mean values (one for experiment) were modelled according to the Gompertz equation modified by Zwietering et al. [30]:

$$y = k + A * exp\{-exp[(\mu_{max} * e/A) * (\lambda - t) + 1]\}, \quad (1)$$

where y is the concentration of the microorganism (Log CFU/ g), k is the initial level of the dependent variable to be modelled, A is the difference between the decimal logarithm of the initial value of cell concentration and the decimal logarithm of maximum bacteria growth attained at the stationary phase (Log CFU/g), μ_{max} is the maximal growth rate (1/day), λ the lag time (day) and t the time.

Following Castillejo Rodriguez et al. [31], the sanitary risk time for the growth of *L. monocytogenes* in our samples was determined as the time (in days) that it took to observe an increase of 2 Log CFU/g of the count of this microorganism in food as follow:

$$\text{sanitary risk time [SRT]} = 2/\mu, \quad (2)$$

where μ is the maximal growth rate.

For the growth of *Ps. fluorescens*, the maximum acceleration of microbial growth (dy^2/dt^2 (day)), known as stability time, was also estimated with the Gompertz equation, following Riva et al. [32].

To determine whether significant differences ($p < 0.05$) existed among the parameters calculated by using the Gompertz equation, a one-way analysis of variance (ANOVA), followed by Tukey's test, was conducted.

Modeling was performed through the software Statistica for Windows version 10.0 (Statsoft, Tulsa, OK, USA).

3. Results and Discussion

3.1. Effect of Probiotic Biofilms on Pathogen Sessile Growth

In order to evaluate the effect of probiotic biofilms on the development of pathogenic microorganisms, evidence was provided on the growth in sessile form of *L. monocytogenes*, *E. coli* O157:H7, *St. aureus* and *S. enterica*. Table 2 shows the cellular loads in sessile form relating to the targets studied; the data analysis shows how the pathogens studied were able to develop in all samples, even if they exhibited a wide range in their ability to colonize the surface, with the highest initial adhesion recovered for *S. enterica* (about 6 log CFU/cm^2) against the lowest one (about 4 log CFU/cm^2) recovered for *L. monocytogenes*. However, cellular loads were always lower in ACT samples (presence of probiotic biofilm, about 6.5–7 log CFU/cm^2) compared to the CNT samples (absence of probiotic biofilm), highlighting that the studied biofilm was able to control the growth of all inoculated pathogenic targets. To quantify the effectiveness of probiotic biofilms in slowing down the pathogens' adhesion, for each time of analysis the difference between the cellular loads recovered in CNT and ACT samples was calculated. As it can be inferred from Table 2, for *E. coli* O157:H7, there was a significant decrease in cell load compared to control of more than 1 and 2 logarithmic cycles after 4 and 48 h of incubation, respectively, and the biofilm efficacy increased over time. Similar results were observed for *St. aureus*. On the contrary, for *L. monocytogenes* the effectiveness of probiotic biofilm was maximum after 4 h (1.43 ± 0.28), but it decreased over time; this loss of efficacy was also recorded for *S. enterica*, with cell load reductions ranging from 1 to 0.2 logarithmic cycles after 24 and 48 h, respectively. As expected, biofilms were odorless and invisible to the naked eye. The idea to use probiotics into the prevention of infections and other diseases has already been proposed [7], and is also stimulated by the need of new alternative intervention strategies to combat bacteria pathogenesis due to the increasing evidence of antibiotics resistance of many pathogens. Abdelhamid et al. [33] observed that cell-free preparations of different probiotics belonging to *Lactobacillus* and *Bifidobacterium* species were able to reduce the growth of *E. coli*, whereas Kaboosi [34] showed that probiotics from yogurts had antibacterial effects against Gram negative bacteria such as *E. coli*, *Salmonella* Typhi and *Ps. aeruginosa*, and Gram positive bacteria such as *S. aureus*. Similarly, Tejero-Sariñena et al. [35] found that 15 strains of probiotics had antibacterial properties against gram negative *Salmonella* Typhimurium and *Clostridium difficile*. However, most of these studies propose the use of compounds (mainly bio-surfactants) with antimicrobial activity produced by probiotics, and contained in their cell-free supernatants [20–27]; on the contrary, this study proposes the use of a probiotic biofilm that exploits the in vivo metabolism of two bacterial strains (*Lactobacillus* and *Bifidobacterium*) adhering on abiotic surfaces and not the substances secreted by probiotics and subsequently recovered and used. In 2014 Schobitz et al. [36] proposed a biocontroller consisting of the thermally treated fermentate (TTF) from two *Carnobacterium maltaromaticum* strains (ATCC PTA 9380 and ATCC PTA 9381), a strain of *Enterococcus mundtii* (ATCC PTA 9382), plus nisin at a concentration of 1000 IU/mL, with all these components entrapped in an alginate matrix supported by a mesh-type fabric. The strains used in our study are different, and no bacteriocin or polymer is used, but the proposed probiotic biofilm should be formed on different surfaces chosen according the purpose (an active packaging and/or a medical device). Moreover, our solution, thanks to the maintenance of a continuous metabolism, should ensure an uninterrupted and stronger activity of the active substances (mainly bacteriocins and/or other LAB-produced antimicrobial compounds such as hydrogen peroxide, carbon dioxide, diacetyl, organic acids), being the same in loco produced [37]. Similar to our study, Gomez et al. [28] used in situ biofilms formed by potential probiotic LAB strains isolated from Brazilian's foods (*Lactococcus lactis* VB69, *L. lactis* VB94, *Lactobacillus sakei* MBSa1, *Lactobacillus curvatus* MBSa3, *L. lactis* 368, *Lactobacillus helveticus* 354, *Lactobacillus casei* 40, and *Weissela viridescens* 113) to inhibit pathogenic growth: they found the total inhibition in pathogens *E. coli* O157:H7, *L. monocytogenes* and *Salmonella* Typhimurium biofilm formation, in 24, 48, and 72 h of exposure using *L. lactis* 368, *Lactobacillus curvatus* MBSa3 and *Lactobacillus sakei* MBSa1. For the other strains, the inhibition was time-dependent and varied according to the strain and target pathogen; for *L. monocytogenes*, reductions ranged from 4- to 7-log units over 24 and 48 h, and the inhibition was observed only within the first 24–48 h, after which the pathogen was able to grow. In *Salmonella* Typhimurium and *E. coli* O157:H7 experiments, sessile cells were not detected during 24 h of incubation in the presence of most LAB tested; during 48 and 72 h, reductions between 5 and 3 log for *E. coli* O157:H7 and 4 log for *Salmonella* Typhimurium were achieved.

Table 2. Cellular loads (Log CFU/cm^2) recovered for *Listeria monocytogenes*, *Escherichia coli* O157:H7, *Staphylococcus aureus* and *Salmonella enterica* during their sessile growth with (ACTIVE, ACT) or without (CONTROL, CNT) probiotic biofilms.

	L. monocytogenes		
Time (h)	**CNT**	**ACT**	**** Biofilm Efficacy**
0	4.21 ± 0.01 [A,*]	3.46 ± 0.12 [B]	0.75 ± 0.17 [a,***]
4	4.82 ± 0.16 [A]	3.39 ± 0.20 [B]	1.43 ± 0.28 [b]
24	4.83 ± 0.13 [A]	4.10 ± 0.10 [B]	0.73 ± 0.14 [a]
30	5.18 ± 0.25 [A]	4.31 ± 0.11 [B]	0.87 ± 0.16 [a]
48	4.91 ± 0.01 [A]	4.23 ± 0.10 [B]	0.68 ± 0.14 [a]
	E. coli O157:H7		
Time (h)	**CNT**	**ACT**	**Biofilm Efficacy**
0	5.49 ± 0.01 [A]	5.20 ± 0.25 [A]	0.29 ± 0.35 [a]
4	5.43 ± 0.14 [A]	4.19 ± 0.22 [B]	1.24 ± 0.31 [b]
24	5.56 ± 0.41 [A]	4.10 ± 0.01 [B]	1.46 ± 0.01 [b]
30	6.13 ± 0.30 [A]	3.82 ± 0.01 [B]	2.31 ± 0.01 [c]
48	6.00 ± 0.25 [A]	3.80 ± 0.20 [B]	2.20 ± 0.28 [c]
	St. aureus		
Time (h)	**CNT**	**ACT**	**Biofilm Efficacy**
0	5.07 ± 0.20 [A]	4.88 ± 0.01 [A]	0.19 ± 0.01 [a,b]
4	5.02 ± 0.03 [A]	4.70 ± 0.33 [A]	0.32 ± 0.47 [b,c]
24	5.57 ± 0.14 [A]	4.89 ± 0.19 [B]	0.68 ± 0.27 [b,c]
30	5.16 ± 0.01 [A]	3.86 ± 0.22 [B]	1.30 ± 0.31 [c,d]
48	5.16 ± 0.30 [A]	3.71 ± 0.05 [B]	1.45 ± 0.07 [d]

Table 2. *Cont.*

Salmonella enterica			
Time (h)	**CNT**	**ACT**	**Biofilm Efficacy**
0	5.94 ± 0.10 ^{A}	4.37± 0.10 ^{B}	1.57 ± 0.14 ^{a}
4	5.38 ± 0.10 ^{A}	4.47 ± 0.16 ^{B}	0.91 ± 0.23 ^{b}
24	5.53 ± 0.15 ^{A}	4.54 ± 0.13 ^{B}	0.99 ± 0.18 ^{b}
30	5.35 ± 0.23 ^{A}	4.88 ± 0.05 ^{B}	0.47 ± 0.07 ^{c}
48	4.98 ± 0.30 ^{A}	4.77 ± 0.00 ^{B}	0.21 ± 0.00 ^{c}

* A, B, Values in the same lines with different letters are significantly different (Student's t-test) ($p < 0.05$). ** Biofilm Efficacy = CNT–ACT. *** a, b, c, d, Values in the same columns with different letters are significantly different (one-way ANOVA and Tukey's test) ($p < 0.05$).

3.2. Application as Potential Active Packaging

Once ascertained the effects on pathogens growth, the research focused on the formation of the probiotic biofilm on different packaging materials, in order to individuate an innovative packaging system.

The results obtained are shown in Table 3; after only 2 h, the probiotic biofilm was successfully formed on all tested materials, except for waxed paper. The sessile cellular load ranged from 5.77 log CFU/cm^2 (grease-proof paper) to 6.94 log CFU/cm^2 (polyethylene). After 96 h, polyethylene and ceramic resulted the materials on which the highest adhesion was recorded (6.54 log CFU/cm^2). In general, any surface (plastic, rubber, glass, metal, paper, cement, stainless steel or wood, or food products themselves) are vulnerable to biofilm development and each biofilm is different, thus suggesting that every situation should be analysed individually and specifically [38].

Table 3. Cellular probiotic load in sessile form (log CFU/cm^2) observed on common packaging materials used in the food industry and on ceramic.

Materials	**Cellular Probiotic Load in Sessile Form (log CFU/cm^2)**		
	2 h	**24 h**	**96 h**
Polypropylene (PP)	6.64 ± 0.00 ^{A}	6.14 ± 0.48 ^{A}	5.88 ± 0.23 ^{A}
Polyvinyl chloride (PVC)	6.54 ± 0.14 ^{A}	5.65 ± 0.10 ^{A}	5.87 ± 0.30 ^{A}
Greaseproof paper (GP)	5.77 ± 0.23 ^{B}	5.24 ± 0.15 ^{B}	5.25 ± 0.06 ^{B}
Waxed paper (WP)	No adhesion	4.61 ± 0.22 ^{C}	4.53 ± 0.13 ^{C}
Polyethylene (PE)	6.94 ± 0.00 ^{A}	6.03 ± 0.38 ^{A}	6.54 ± 0.14 ^{D}
Ceramic	6.86 ± 0.20 ^{A}	6.24 ± 0.23 ^{A}	6.54 ± 0.14 ^{D}

A, B, C, Values in the same columns with different letters are significantly different (one-way ANOVA and Tukey's test) ($p < 0.05$).

Once individuated in polyethylene (PE) the material able to ensure the greatest adhesion of probiotics, in a second step, the attention was focused only on this material and it was used to test the potential for probiotic biofilms to control the growth of microorganisms in soft cheeses. The products were inoculated with *L. monocytogenes* (challenge test A) and *Ps. fluorescens* (challenge test B), wrapped in PE pellicles with pre-formed probiotic biofim, packed both in air and under vacuum, and stored at 4 and 15 °C. These model bacteria were chosen as main representatives of pathogen and spoilage bacteria naturally contaminating soft cheese [39,40].

At 4 °C, the cellular load of *L. monocytogenes* remained lower than 3 log CFU/g for the entire observation period (28 days), regardless the presence of the probiotic biofilm or the packaging. On the other hand, at 15 °C (simulated thermal abuse), the λ length was always longer in samples containing probiotic biofilms (EXP samples), if compared to CNT samples (without probiotic biofilms) (Table 4): its value increased from 0.04 to 3.37 days (in air packaging, Figure 1A) and from 0.00 to 2.40 days (under vacuum, Figure 1B). The growth rate (μ_{max}) was also influenced by the presence of probiotic biofilms, recording a decrease from about 0.7 to 0.4 Log(CFU/g)/day, in both packaging conditions. The maximum cell load reached in the stationary phase (A + N_0) was not influenced, reaching approximately 5.6–5.7 log CFU/g, regardless of the presence or absence of probiotic biofilms. The cellular load of lactic bacteria (LAB) was also monitored, as well as pH and water activity. At 4 °C, the initial LAB count was 5.75 ± 0.18 log CFU/g in the control samples against 8.32 ± 0.20 log CFU/g in the experimental cheeses; after 28 days, there were no statistically significant differences between the samples (regardless of the presence of probiotic biofilms and the type of packaging), recording cellular loads between 7 and 8 log CFU/g (data not shown).

Table 4. Kinetic parameters calculated by fitting Gompertz equation to the experimental data by *L. monocytogenes* and *Ps. fluorescens* during their growth in soft cheeses with (EXP) or without (CNT) probiotic biofilms, packed in AIR o under vacuum (UV) and stored at 15 °C. (A + No) is the maximum bacterial load attained at the stationary phase, μ_{max} is the maximal growth rate, λ is the lag time, TRS is the sanitary risk time, ST (stability time) is the maximum acceleration of microbial growth.

L. monocytogenes				
	A + No [Log CFU/g]	**μmax [Log(CFU/g)/day]**	**λ [day]**	**TRS * [day]**
CNT AIR	5.66 ± 0.31 ^{A}	0.69 ± 0.14 ^{A}	0.04 ± 0.67 ^{A}	2.88
EXP AIR	5.37 ± 0.20 ^{A}	0.43 ± 0.19 ^{A}	3.37 ± 1.06 ^{B}	4.62
CNT UV	5.94 ± 0.85 ^{A}	0.68 ± 0.09 ^{A}	0.00 ± 0.00 ^{A}	2.95
EXP UV	5.39 ± 0.11 ^{A}	0.47 ± 0.09 ^{A}	2.40 ± 0.72 ^{B}	4.30
Ps. fluorescens				
	A + No [Log CFU/g]	**μmax [Log(CFU/g)/day]**	**λ [day]**	**ST ** [day]**
CNT AIR	5.95 ± 0.33 ^{A}	0.41 ± 0.23 ^{A}	0.54 ± 0.61 ^{A}	2.54
EXP AIR	5.66 ± 0.18 ^{A}	0.33± 0.09 ^{A}	4.40 ± 0.74 ^{B}	6.59
CNT UV	6.00 ± 0.19 ^{A}	0.28 ± 0.05 ^{A}	0.00 ± 0.50 ^{A}	3.03
EXP UV	5.84 ± 0.25 ^{A}	0.26 ± 0.05 ^{A}	3.33 ± 0.76 ^{B}	6.35

A, B, Values in the same columns with different letters are significantly different (one-way ANOVA and Tukey's test) ($p < 0.05$). *, TRS, sanitary risk time, i.e., the time required (in days) to observe an increase of 2 log CFU/g in *L. monocytogenes* count [30]. **, stability time, i.e., the maximum acceleration of microbial growth [dy^2/dt^2 (day)] [31].

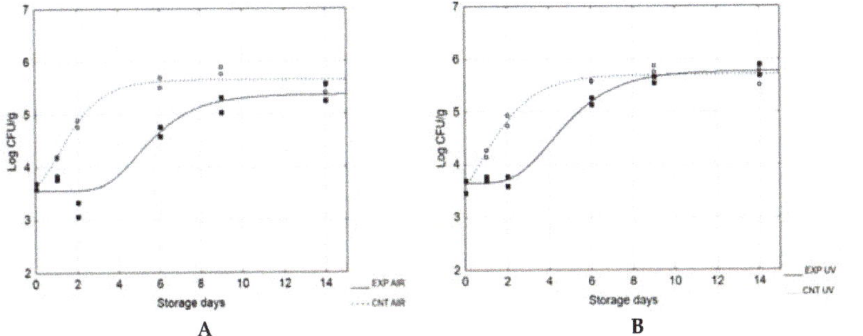

Figure 1. Evolution of *L. monocytogenes* during the challenge test at 15 °C. EXP, cheeses stored with probiotic biofilms; CNT, cheeses stored without probiotic biofilm. (**A**), AIR packaging; (**B**), under vacuum packaging (UV).

Additionally, for the pH, no significant differences between the samples were observed; this parameter decreased from 5.31–5.39 to 4.70–4.84 at the end of storage. In all samples, the value of water activity remained constant (0.99–1.00) for the entire duration of the experimentation (data not shown). Similar results were observed at 15 °C.

During the experimentation, both at 4 and 15 °C, a gradual decrease of the score from 10 to about 5.5–6 was recorded (end of storage), regardless of the presence of probiotic biofilms and the type of packaging applied, showing that the probiotic microorganisms had no impact on the sensory characteristics of cheeses; as an example, Figure 2 shows the sensorial scores for colour, odour, texture and overall acceptability of cheeses recovered during storage at 4 °C.

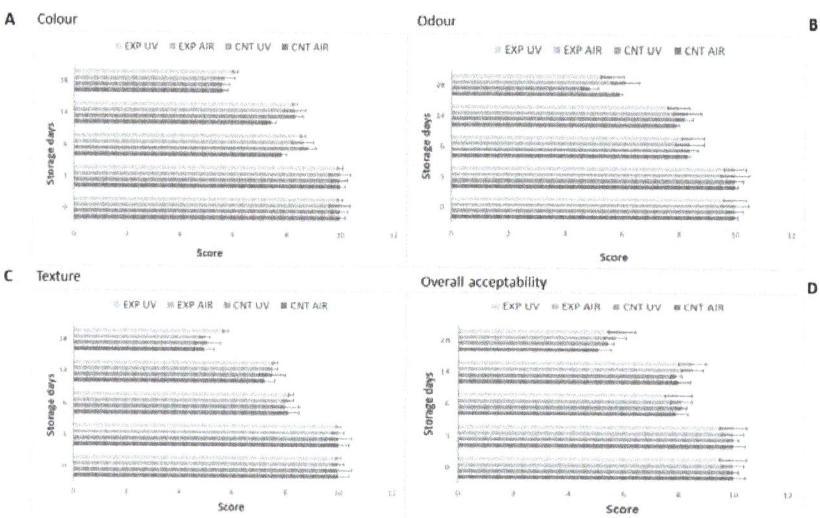

Figure 2. Sensorial scores for colour (**A**), odour (**B**), texture (**C**) and overall acceptability (**D**) of cheeses inoculated with *L. monocytogenes* stored at 4 °C. Mean values ± standard deviation. EXP, cheeses stored with probiotic biofilms; CNT, cheeses stored without probiotic biofilm.

Figure 3; Figure 4 show the evolution of *Ps. fluorescens* during the growth on EXP and CNT cheeses stored at 4 and 15 °C, respectively. The target microorganism was able to grow under all tested conditions, regardless of the presence of probiotic biofilms and the type of packaging. At 4 °C (Figure 3), the presence of the probiotic biofilm was able to influence the maximum cellular load reached in the stationary phase ($A + N_0$), which was significantly lower in the EXP samples (5.59–5.72 log CFU/ g) compared to the CNT samples (6.36–6.39 log CFU/g). No influence was observed about the λ length and the maximum growth rate (μ_{max}).

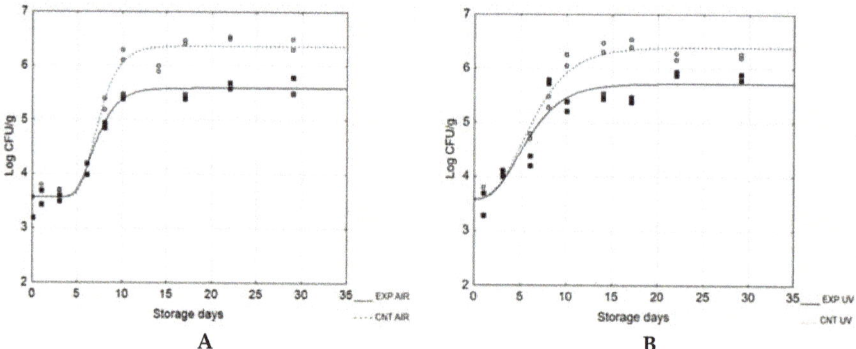

Figure 3. Evolution of *Ps. fluorescens* during the challenge test at 4 °C. EXP, cheeses stored with probiotic biofilms; CNT, cheeses stored without probiotic biofilm. (**A**), AIR packaging; (**B**), under vacuum packaging (UV).

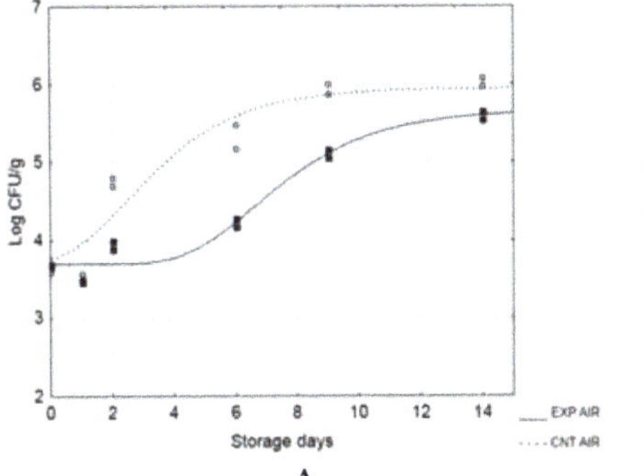

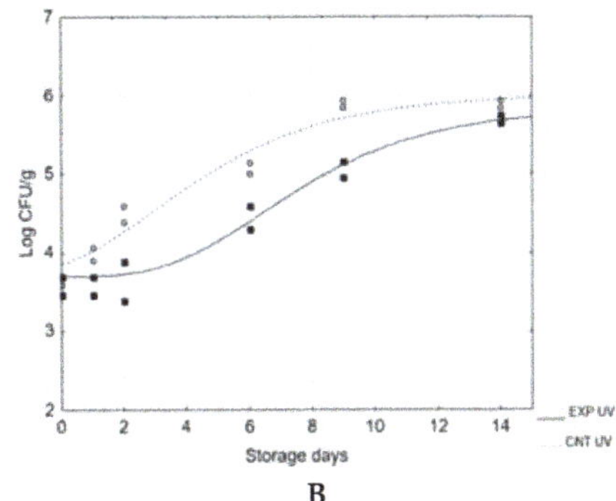

Figure 4. Evolution of *Ps. fluorescens* during the challenge test at 15 °C. EXP, cheeses stored with probiotic biofilms; CNT, cheeses stored without probiotic biofilm. (**A**), AIR packaging; (**B**), under vacuum packaging (UV).

During storage at 15 °C (Figure 4), the presence of probiotic biofilms significantly slowed the growth of the target microorganism: λ increased from 0.54 to 4.40 days and from 0.01 to 3.30 days, in air and vacuum packaging, respectively. The maximum growth rate and the maximum cell load reached in the stationary phase were also lower in the EXP samples (probiotic biofilms) than the control samples, regardless of the packaging applied.

At both 4 and 15 °C, data on LAB, pH and water activity were similar to those observed in the challenge test with *L. monocytogenes* (data not shown). Results of sensory analyses confirmed that the probiotic microorganisms had no impact on the organoleptic characteristics of cheeses (data not shown).

To highlight the effectiveness of probiotic biofilms to slow the decay of the microbiological quality of soft cheeses at 15 °C, Table 4 shows the kinetic parameters of Gompertz equation accompained by two other parameters (TRS and stability time). In a well-known study on the growth of *L. monocytogenes* in food, Castillejo Rodriguez et al. [30] have proposed the sanitary risk time (TRS) for this pathogen as the time required (in days) to observe an increase of 2 log CFU/g in its count, considering that, under normal conditions, such a microorganism is present in foods in very low concentrations. As can be seen, for soft cheeses wrapped in probiotic biofilms and packaged both in air and under vacuum, the TRS was equal to 4.40–4.60 days; on the contrary, the same methods of packaging, applied to the control samples, allowed *L. monocytogenes* to reach risky cell counts in shorter times (2.88–2.95 days) ($p < 0.05$).

For the tests conducted with *Ps. fluorescens* at 15 °C, Table 4 also shows the stability time [31,41] which represents the maximum acceleration of microbial growth and indicates how long the product remains stable: after this time, an irreversible decay of the product begins. This parameter is generally used as an alternative to shelf life: the underlying principle implies that microbial degradation has to show a rate of the same order of magnitude as at the shelf life zero time. This condition is no longer met when microbial growth attains its maximum acceleration, because beyond such a threshold the system undergoes very fast changes with a rapid loss of the generally accepted safety or quality requirements. This principle seems more reliable than the current practice that defines food stability according to the ratio between attained and starting microbial population levels. The stability time increased by more than 3 and 4 days, in vacuum packaging and in air, respectively, highlighting the effectiveness of biofilms in slowing the decay of the microbiological quality of soft cheese.

Regarding the inhibitory effect of LAB against *L. monocytogenes*, some studies have already explored the possibility to use a preformed biofilm to inhibit the pathogen growth [15,16,18,42]. Namely, Guerrieri et al. [15] showed the potential of a *Lactobacillus plantarum* strain to reduce the pathogen growth over a 10-day period (about 4-log reduction). Mariani et al. [42] used the native biofilm microflora of wooden cheese ripening shelves to achieve a 1- to 2-log reduction over a 12-day period. In previous studies, we have evaluated the use of LAB biofilms as a means to control the growth of *L. monocytogenes* in soft cheeses [16] and in laboratory media [18], finding that sessile LAB biofilms were able to delay the growth of *L. monocytogenes*. An anti-listerial activity was also observed by Léonard et al. [43] during their studies on biopolymeric matrices based on alginate and alginate-caseinate (an aqueous two-phase system) entrapping *Lactococcus lactis* subsp. *lactis* LAB3 cells and by Barbosa et al. [44] who entrapped *Lactobacillus curvatus* in calcium alginate: the effect against the pathogen was correlated to antimicrobial metabolites of proteinaceous nature.

4. Conclusions

This study has explored whether probiotic bacteria able to adhere on different surfaces (i.e., packaging materials, ceramic, plastic, paper, polymers, etc.) could be used as new biotechnological solutions for industrial applications by biocontrolling the growth of pathogenic and spoilage bacteria.

The results obtained have shown the studied biofilm was able to delay the growth of some pathogenic targets; in fact, cellular pathogenic loads were always lower in presence of probiotic biofilm compared to its absence. For *E. coli* O157:H7, a significant cell load decrease (>1–2 logarithmic cycles) was recorded, whereas for *L. monocytogenes*, *St. aureus* and *S. enterica*, cell load reductions ranged from 0.5 to 1.5 logarithmic cycles.

After only 2 h, the probiotic biofilm was successfully formed on polypropylene, polyvinyl chloride, greaseproof paper, polyethylene and ceramic, with polyethylene and ceramic resultingly being the material with the highest adhesion (6.54 log CFU/cm^2). When testing as a tool to control the growth of microorganisms in soft cheeses, the results highlighted the effectiveness of biofilms in slowing the growth of *L. monocytogenes* by prolonging their microbiological stability at 15 °C by more than 3 and 4 days.

The results obtained suggest that the developed probiotic lactic acid bacteria biofilms have a good potential to be used as biocontrol agents against pathogenic and food spoilage bacteria through exclusion mechanisms: however, the mechanisms responsible for the inhibition have to be deeply investigated.

5. Patents

B.S., A.L. and M.R.C. applied for a patent covering the use of probiotic biofilms as a mean to control pathogens growth: Method for producing microbial probiotic biofilms and uses thereof (WO2017203440). International Application No.: PCT/IB2017/053055. National Application No.: P1287IT.

Author Contributions: Conceptualization, B.S, A.L. and M.R.C.; methodology, B.S. and M.R.C.; investigation, B.S.; resources, M.R.C.; writing—original draft preparation, B.S, A.L.and M.R.C.; writing—review and editing, B.S, A.L., V.R. and M.R.C. All authors have read and agreed to the published version of the manuscript.

Funding: This research received no external funding.

Conflicts of Interest: The authors applied for a patent covering the use of probiotic biofilms as a mean to control pathogens growth.

References

1. Ouwehand, A.C.; Salminen, S.; Isolauri, E. Probiotics: An overview of beneficial effects. *Antonie Van Leeuwenhoek* **2002**, *82*, 279–289. [CrossRef]
2. Hill, C.; Guarner, F.; Reid, G.; Gibson, G.R.; Merenstein, D.J.; Pot, B.; Morelli, L.; Canani, R.B.; Flint, H.J.; Salminen, S.; et al. Expert consensus document: The international scientific association for probiotics and prebiotics consensus statement on the scope and appropriate use of the term probiotic. *Nat. Rev. Gastroenterol. Hepato.* **2014**, *11*, 506–514. [CrossRef]
3. Charalampopoulos, D.; Rastall, R.A. *Prebiotics and Probiotics Science and Technology*; Springer: New York, NY, USA, 2009; p. 1262.

4. Corbo, M.R.; Speranza, B.; Sinigaglia, M. Microbial biofilms in the food industry: only negative implications? In *Biofilms: Formation, Development and Properties*; Bailey, W.C., Ed.; Nova Science Publishers: New York, NY, USA, 2009; pp. 585–594.
5. Carducci, A.; Verani, M.; Lombardi, R.; Casini, B.; Privitera, G. Environmental survey to assess viral contamination of air and surfaces in hospital settings. *J. Hosp. Infect.* **2011**, *77*, 242–247. [CrossRef] [PubMed]
6. Eschenbach, D.A.; Patton, D.L.; Hooton, T.M.; Meier, A.S.; Stapleton, A.; Aura, J.; Agnew, K. Effects of vaginal intercourse with and without a condom on vaginal flora and vaginal epithelium. *J. Infect. Dis.* **2001**, *183*, 913–918. [CrossRef] [PubMed]
7. Vuotto, C.; Barbanti, F.; Mastrantonio, P.; Donelli, G. *Lactobacillus brevis* CD2 inhibits *Prevotella melaninogenica* biofilm. *Oral. Dis.* **2014**, *20*, 668–674. [CrossRef] [PubMed]
8. Ma, L.; Lv, Z.; Su, J.; Wang, J.; Yan, D.; Wei, J.; Pei, S. Consistent Condom Use Increases the Colonization of *Lactobacillus crispatus* in the Vagina. *PLoS ONE* **2013**, *8*, e70716. [CrossRef] [PubMed]
9. Leccese Terraf, M.C.; Juarez Tomas, M.S.; Nader-Macıas, M.E.F.; Silva, C. Screening of biofilm formation by beneficial vaginal lactobacilli and influence of culture media components. *J. Appl. Microbiol.* **2012**, *113*, 1517–1529. [CrossRef] [PubMed]
10. Salo, S.; Ehavald, H.; Raaska, L.; Vokk, R.; Wirtanen, G. Microbial surveys in Estonian dairies. *LWT Food Sci. Technol.* **2006**, *39*, 460–471. [CrossRef]
11. Sharma, M.; Anand, S.K. Biofilms evaluation as an essential component of HACCP for food/dairy processing industry e a case. *Food Control* **2002**, *13*, 469–477. [CrossRef]
12. Waak, E.; Tham, W.; Danielsson-Tham, M.L. Prevalence and fingerprinting of *Listeria monocytogenes* strains isolated from raw whole milk in farm bulk tanks and in dairy plant receiving tanks. *Appl. Env. Microbiol.* **2002**, *68*, 3366–3370. [CrossRef]
13. Speranza, B.; Corbo, M.R. The impact of biofilms on food spoilage. In *The Microbiological Quality of Food: Foodborne Spoilers*; Bevilacqua, A., Corbo, M.R., Sinigaglia, M., Eds.; Elsevier: Oxford, UK, 2017; pp. 259–282.
14. Guillier, L.; Stahl, V.; Hezard, B.; Notz, E.; Briandet, R. Modelling the competitive growth between *Listeria monocytogenes* and biofilm microflora of smear cheese wooden shelves. *Int. J. Food Microbiol.* **2008**, *128*, 51–57. [CrossRef]
15. Guerrieri, E.; de Niederhäusern, S.; Messi, P.; Sabia, C.; Iseppi, R.; Anacarso, I.; Bondi, M. Use of lactic acid bacteria (LAB) biofilms for the control of *Listeria monocytogenes* in a small-scale model. *Food Control* **2009**, *20*, 861–865. [CrossRef]
16. Speranza, B.; Sinigaglia, M.; Corbo, M.R. Non starter lactic acid bacteria biofilms: A means to control the growth of *Listeria monocytogenes* in soft cheese. *Food Control* **2009**, *20*, 1063–1067. [CrossRef]
17. Vandini, A.; Temmerman, R.; Frabetti, A.; Caselli, E.; Antonioli, P.; Balboni, P.G.; Platano, D.; Branchini, A.; Mazzacane, S. Hard Surface Biocontrol in Hospitals Using Microbial-Based Cleaning Products. *PLoS ONE* **2014**, *9*, e108598. [CrossRef]
18. Speranza, B.; Liso, A.; Corbo, M.R. Use of design of experiments to optimize the production of microbial probiotic biofilms. *PeerJ* **2018**, *6*, e4826. [CrossRef]
19. Method for Producing Microbial Probiotic Biofilms and Uses thereof WO2017203440. Available online: https://patents.google.com/patent/WO2017203440A1/en (accessed on 25 November 2019).
20. Gudiña, E.J.; Teixeira, J.A.; Rodrigues, L.R. Isolation and functional characterization of a biosurfactant produced by *Lactobacillus paracasei*. *Colloid Surf. B* **2010**, *76*, 298–304. [CrossRef]
21. Jones, S.E.; Versalovic, J. Probiotic *Lactobacillus reuteri* biofilms produce antimicrobial and anti-inflammatory factors. *BMC Microbiol* **2009**, *9*, 35. [CrossRef]
22. Rodrigues, L.; Banat, I.M.; Teixeira, J.; Oliveira, R. Biosurfactants: potential applications in medicine. *J. Antimicrob. Chemoth.* **2006**, *57*, 609–618. [CrossRef]
23. Rodrigues, L.; Banat, I.M.; Teixeira, J.; Oliveira, R. Strategies for the prevention of microbial biofilm formation on silicone rubber voice prostheses. *J. Biomed Mater Res. B* **2007**, *81*, 358–370. [CrossRef]
24. Rodrigues, L.; van der Mei, H.C.; Banat, I.M.; Teixeira, J.; Oliveira, R. Inhibition of microbial adhesion to silicone rubber treated with biosurfactant from *Streptococcus thermophilus* A. *FEMS Immunol. Med. Microbiol.* **2006**, *46*, 107–112. [CrossRef]
25. Rodrigues, L.; van der Mei, H.C.; Teixeira, J.; Oliveira, R. Influence of biosurfactants from probiotic bacteria on formation of biofilms on voice prostheses. *Appl. Env. Microbiol.* **2004**, *70*, 4408–4410. [CrossRef]

26. Rodrigues, L.R.; Banat, I.M.; van der Mei, H.C.; Teixeira, J.A.; Oliveira, R. Interference in adhesion of bacteria and yeasts isolated from explanted voice prostheses to silicone rubber by rhamnolipid biosurfactants. *J. Appl. Microbiol.* **2006**, *100*, 470–480. [CrossRef]
27. Rodrigues, L.R.; Teixeira, J.A.; van der Mei, H.C.; Oliveira, R. Physicochemical and functional characterization of a biosurfactant produced by *Lactococcus lactis* 53. *Colloid Surf. B* **2006**, *49*, 79–86. [CrossRef]
28. Gómez, N.C.; Ramiro, J.M.P.; Quecan, B.X.V.; de Melo Franco, B.D.G. Use of potential probiotic lactic acid bacteria (LAB) biofilms for the control of *Listeria monocytogenes*, *Salmonella* Typhimurium, and *Escherichia coli* O157: H7 biofilms formation. *Front. Microbiol.* **2016**, *7*, 863. [CrossRef]
29. Mandakhalikar, K.D.; Rahmat, J.N.; Chiong, E.; Neoh, K.G.; Shen, L.; Tambyah, P.A. Extraction and quantification of biofilm bacteria: Method optimized for urinary catheters. *Sci. Rep.* **2018**, *8*, 8069. [CrossRef]
30. Zwietering, M.H.; Jogenburger, I.; Rombouts, F.M.; Van't Riet, K. Modelling of the bacterial growth curve. *Appl. Environ. Microbiol.* **1990**, *56*, 1975–1981. [CrossRef]
31. Castillejo Rodrìguez, A.M.; Barco Alcalà, E.; Garcia Gimeno, R.M.; Zurera Cosano, G. Growth modelling of *Listeria monocytogenes* in packaged fresh green asparagus. *Food Microbiol.* **2000**, *17*, 421–427. [CrossRef]
32. Riva, M.; Franzetti, L.; Galli, A. Microbiological quality and shelf life modeling of ready-to-eat cicorino. *J. Food Prot.* **2001**, *64*, 228–234. [CrossRef]
33. Abdelhamid, A.G.; Esaam, A.; Hazaa, M.M. Cell free preparations of probiotics exerted antibacterial and antibiofilm activities against multidrug resistant *E. coli*. *Saudi Pharm. J.* **2018**, *26*, 603–607. [CrossRef]
34. Kaboosi, H. Antibacterial effects of probiotics isolated from yoghurts against some common bacterial pathogens. *Afr. J. Microbiol. Res.* **2011**, *5*, 4363–4367. [CrossRef]
35. Tejero-Sariñena, S.; Barlow, J.; Costabile, A.; Gibson, G.R.; Rowland, I. Antipathogenic activity of probiotics against *Salmonella Typhimurium* and *Clostridium difficile* in anaerobic batch culture systems: is it due to synergies in probiotic mixtures or the specificity of single strains? *Anaerobe* **2013**, *24*, 60–65. [CrossRef]
36. Barbosa, M.S.; Todorov, S.D.; Jurkiewicz, C.H.; Franco, B.D.G.M. Bacteriocin production by *Lactobacillus curvatus* MBSa2 entrapped in calcium alginate during ripening of salami for control of *Listeria monocytogenes*. *Food Control* **2015**, *47*, 147–153. [CrossRef]
37. Gálvez, A.; Abriouel, H.; Ben Omar, N. Bacteriocin-based strategies for food biopreservation. *Int. J. Food Microbiol.* **2007**, *120*, 51–70. [CrossRef]
38. Srey, S.; Jahid, I.K.; Ha, S. Biofilm formation in food industries: a food safety concern. *Food Control* **2013**, *31*, 572–585. [CrossRef]
39. De Buyser, M.L.; Dufour, B.; Maire, M.; Lafarge, V. Implication of milk and milk products in food-borne diseases in France and different industrialized countries. *Int. J. Food Microbiol.* **2001**, *67*, 1–17. [CrossRef]
40. Martin, N.H.; Murphy, S.C.; Ralyea, R.; Boor, K.J. When cheese gets the blues: *Pseudomonas fluorescens* as the causative agent of cheese spoilage. *J. Dairy Sci.* **2011**, *94*, 3176–3183. [CrossRef]
41. Corbo, M.R.; Altieri, C.; D'Amato, D.; Campaniello, D.; Del Nobile, M.A.; Sinigaglia, M. Effect of temperature on shelf life and microbial population of lightly processed cactus pear fruit. *Postharvest Biol. Technol.* **2004**, *31*, 93–104. [CrossRef]
42. Mariani, C.; Oulahal, N.; Chamba, J.F.; Dubois-Brissonnet, F.; Notz, E.; Briandet, R. Inhibition of *Listeria monocytogenes* by resident biofilms present on wooden shelves used for cheese ripening. *Food Control* **2011**, *22*, 1357–1362. [CrossRef]
43. Léonard, L.; Degraeve, P.; Gharsallaoui, A.; Saurel, R.; Oulahal, N. Design of biopolymeric matrices entrapping bioprotective lactic acid bacteria to control *Listeria monocytogenes* growth: Comparison of alginate and alginate-caseinate matrices entrapping *Lactococcus lactis* subsp. *lactis cells*. *Food Control* **2014**, *37*, 200–209. [CrossRef]
44. Schöbitz, R.; González, C.; Villarreal, K.; Horzella, M.; Nahuelquín, Y.; Fuentes, R. A biocontroller to eliminate *Listeria monocytogenes* from the food processing environment. *Food Control* **2014**, *36*, 217–223. [CrossRef]

Article

Probiotic Bifunctionality of *Bacillus subtilis*—Rescuing Lactic Acid Bacteria from Desiccation and Antagonizing Pathogenic *Staphylococcus aureus*

Hadar Kimelman [1,2] and Moshe Shemesh [1,*]

1 Department of Food Sciences, Institute for Postharvest Technology and Food Sciences, Agricultural Research Organization (ARO), Volcani Center, Rishon LeZion 7528809, Israel; hadar.kimelman@mail.huji.ac.il
2 Institute of Dental Sciences, Faculty of Dental Medicine, Hebrew University-Hadassah, Jerusalem 9112102, Israel
* Correspondence: moshesh@agri.gov.il; Tel.: +972-3-9683868

Received: 29 August 2019; Accepted: 23 September 2019; Published: 29 September 2019

Abstract: Live probiotic bacteria obtained with food are thought to have beneficial effects on a mammalian host, including their ability to reduce intestinal colonization by pathogens. To ensure the beneficial effects, the probiotic cells must survive processing and storage of food, its passage through the upper gastrointestinal tract (GIT), and subsequent chemical ingestion processes until they reach their target organ. However, there is considerable loss of viability of the probiotic bacteria during the drying process, in the acidic conditions of the stomach, and in the high bile concentration in the small intestine. *Bacillus subtilis*, a spore-forming probiotic bacterium, can effectively maintain a favorable balance of microflora in the GIT. *B. subtilis* produces a protective extracellular matrix (ECM), which is shared with other probiotic bacteria; thus, it was suggested that this ECM could potentially protect an entire community of probiotic cells against unfavorable environmental conditions. Consequently, a biofilm-based bio-coating system was developed that would enable a mutual growth of *B. subtilis* with different lactic acid bacteria (LAB) through increasing the ECM production. Results of the study demonstrate a significant increase in the survivability of the bio-coated LAB cells during the desiccation process and passage through the acidic environment. Thus, it provides evidence about the ability of *B. subtilis* in rescuing the desiccation-sensitive LAB, for instance, *Lactobacillus rhamnosus*, from complete eradication. Furthermore, this study demonstrates the antagonistic potential of the mutual probiotic system against pathogenic bacteria such as *Staphylococcus aureus*. The data show that the cells of *B. subtilis* possess robust anti-biofilm activity against *S. aureus* through activating the antimicrobial lipopeptide production pathway.

Keywords: beneficial biofilm; bio-coating; *B. subtilis*; probiotics; extracellular matrix; pathogen elimination

---

1. Introduction

Live probiotic microorganisms obtained often with food are thought to improve human health. Thus, probiotics are usually defined as live microbial cells that provide a health benefit to the host when administered in sufficient quantities [1]. Among most prominent probiotic microorganisms are Gram-positive lactic acid bacteria (LAB), which mainly belong to the *Lactobacillus* and *Bifidobacterium* genera [2]. The essential probiotic requirement in terms of the health benefits is a positive influence on the digestion and immune systems [3]. Moreover, probiotics also have a protective role, directly competing with pathogens through signaling interference [4], releasing antimicrobial substances [5] or metabolites such as acids [6–8]. Nevertheless, to exert its beneficial effects, any probiotic organism must survive, establish, and multiply in the host.

Probiotic bacteria are usually delivered as dried cultures, but the process used to prepare them may damage the cell's structure, vitality, and functionality [9,10]. Drying processes involve the removal of a large amount of fluid from the cell, which affects the cellular structure and may, therefore, cause cell death [10,11]. Moreover, probiotic cells must survive shelf life and transit in the gastrointestinal tract, including acid stress in the stomach [11], as well as degradation by enzymes and bile salt in the intestine [12].

One of the main strategies of bacteria to deal with environmental stresses is the formation of a complex structure called a biofilm [13]. In most natural settings, bacteria do not grow as free-living cells but, instead, they form complex polymicrobial structures [14]. The biofilm structures contain less than 10% microorganisms, while the other 90% is the extracellular matrix (ECM) produced by the bacteria themselves. This ECM mainly consists of polysaccharides and other macromolecules such as proteins, enzymes, surfactants, DNA, and lipids [15]. Thus, the biofilm structure is capable of resisting extreme environmental conditions such as transit through the gastrointestinal tract or desiccation [16]. The ECM creates a microenvironment, which might lead to enhanced survival during desiccation [17]. Apparently, hygroscopic polysaccharides are thought to promote biofilm fluidity and resistance to desiccation [18].

Bacillus subtilis, a spore-forming nonpathogenic Gram-positive bacterium, is commonly found in the soil and the gastrointestinal tract (GIT) of some mammals [19]. This bacterium can effectively maintain a favorable balance of microflora in the GIT of the mammalian host [20]. As one of the physiological hallmarks, *B. subtilis* can form a robust biofilm through activation of a dedicated signaling pathway to coordinate expression of genes encoding the ECM [21,22]. Its ECM relies mainly on exopolysaccharides (EPS) synthesized by the *epsA-O* operon and amyloid fibers encoded by the *tasA* located in the *tapA–sipW–tasA* operon [23].

According to recent studies, the use of *Bacillus* species and especially *B. subtilis* as probiotics gained vast interest. Thus, *Bacillus* species were reported to be effective in preventing respiratory infections and gastrointestinal disorders, and overcoming symptoms associated with irritable bowel syndrome [24,25]. However, the mechanism(s) via which *Bacillus* species act as probiotics remains unclear. It appears that the presence of *B. subtilis* helps to maintain a favorable balanced microbiota in the gut and enhances probiotic LAB cell growth and viability [26]. It was also suggested that these probiotic properties are related to its ability to stimulate the immune system [24] and the production of antimicrobial substances [27,28], or even inducing signaling interference against pathogenic microorganisms [4].

Cells of *B. subtilis* produce a vast diversity of antimicrobial substances, amongst them, relatively well-characterized groups of lipopeptides, for instance, surfactins, iturins, and fengycins. These compounds have a wide variety of biological activities such as anti-bacterial, anti-fungal, anti-viral, and anti-tumor activities [29]. They can work in different mechanisms such as disrupting the structure of bacteria members, decreasing the surface and interfacial tension of biofilms [30], and inhibiting quorum sensing, which inhibits biofilm formation [4]. Furthermore, lipopeptides have an essential role in signaling for biofilm formation in *B. subtilis* [31,32].

A model system was recently developed, which enhances biofilm formation by *B. subtilis* through mutual growth with LAB [33]. This system seems to be beneficial for the protection of LAB during heat treatment and through passage in the gastrointestinal tract [33]. The current study presents a further development of the biofilm-based protective coating for probiotic cells via a process defined as a bio-coating. Furthermore, this study provides evidence for two different probiotic functionalities of *B. subtilis*: (i) protecting the LAB during their exposure to desiccation conditions and acid stress; (ii) showing potent anti-microbial activity against pathogenic bacteria such as *Staphylococcus aureus*.

2. Materials and Methods

2.1. Strains and Growth Conditions

Bacterial strains used in this study and their origins are summarized in Table S1 (Supplementary Materials). The lactic acid bacteria (LAB) were routinely grown in either MRS broth (Man, Rogosa & Sharpe) (Hy-labs, Rehovot, Israel) or MRS broth solidified using 1.5% agar (Difco, New-Jersy, USA). In addition, the wild-type (WT) strain NCIB3610 of *B. subtilis* and its derivatives (Table S1, Supplementary Materials) were regularly cultured in Lysogeny broth (LB containing: 10 g of tryptone, 5 g of yeast extract, and 5 g of NaCl per liter) (Difco) or LB solidified with 1.5% agar. Prior to generating starter cultures, LAB and *B. subtilis* cells were grown on the agar-solidified plates for 48 h or overnight, respectively, both at 37 °C. A starter culture of each strain was prepared using a single bacterial colony; the cells of LAB were inoculated into 5 mL of MRS broth overnight without agitation, while the cells of *B. subtilis* were inoculated into LB medium overnight at 30 °C, 150 rpm, until the cultures reached an OD_{600} of approximately 1.5. For co-culture experiments, the modified MRS (MMRS) medium (pH = 7) was used due to its biofilm-promoting capability and its suitability for co-culture cultivation of *B. subtilis* and LAB [33]. Thus, *B. subtilis* cells were mixed with an equal amount of the LAB cells to a final concentration of 10^8 cells/mL of each strain within MMRS. The cells in mixed cultures were incubated aerobically at 37 °C at 50 rpm for 8 h [33]. Cells of *S. aureus* ATCC 25923 were regularly cultured in tryptic soy broth (TSB) (Difco) solidified with 1.5% agar overnight at 37 °C. A starter culture was prepared using a single bacterial colony inoculated into 10 mL of TSB medium overnight at 37 °C, 150 rpm.

2.2. Visualizing Biofilm-Forming Cells Using Confocal Laser Scanning Microscopy (CLSM)

Unlabeled cells of LAB and the CFP-tagged *B. subtilis* cells (YC189) were grown in co-culture into MMRS broth as described above. Cell suspensions of each bacterium grown as mono-species culture served as control samples. One milliliter of each culture was collected and centrifuged at 5000 rpm for 2 min. After removing supernatant, the cells were washed with 1 mL of DW (distilled water) and then the following centrifugation (at 5000 rpm for 2 min) re-suspended in 100 µL of DW. A suspension of 5 µL from each sample was placed on a microscopy glass slide and visualized in a transmitted light microscope using Nomarski differential interference contrast (DIC) and 458-nm laser for CFP excitation (Leica, Wetzler, Germany).

2.3. Growth Curve Analysis of Lab During Growth in Co-Culture

Initially, *B. subtilis* and LAB cells were grown overnight in either LB or MRS as described above. Afterward, the cells were introduced into the MMRS medium, and the co-culture was incubated for 8 h at 37 °C at 150 rpm. Mono-species cultures of LAB and *B. subtilis* were used as control. Every 2 h, 1 mL of each sample was collected for quantification of bacteria by the colony forming unit (CFU) counting method on either MRS (for LAB) or LB (for *B. subtilis*) agar plates. The plates were incubated aerobically at 37 °C for either 48 h (in case of LAB) or 24 h (for *B. subtilis*).

2.4. Analysis of Survival Rates Following Desiccation Treatment

The co-culture samples generated as described above were grown for 8 h aerobically at 37 °C and 50 rpm. The LAB cells grown as a mono-culture were used as a control. One milliliter of each sample was harvested by centrifugation at 4000× *g* for 10 min, washed once with DW, and 50 µL of the sample was placed into wells of a 96-well polystyrene plate (Greiner Bio-One, Monroe, North Carolina, USA). The plate was left open in a dry cabinet (MRC, Holon, Israel), at 40% relative humidity and 25 °C for 20 to 40 h. The MRS agar plates were used to determine the cell counts before drying. For analysis of the viable cell counts following desiccation, 100 µL of DW was added to each well, and the plates were incubated for 5 min at room temperature, before resuspension of the samples by pipetting. Bacterial suspensions were serially diluted and underwent CFU counting on MRS agar plates. The CFU counts were recorded following incubation for 48 h at 37 °C.

2.5. Visualizing Co-Culture Biofilm Following Desiccation Treatment Using Scanning Electron Microscopy (SEM)

The co-culture and mono-culture samples were generated as described above. One milliliter of each sample was harvested by centrifugation (at 4000× *g* for 10 min) and washed once with DW. Then, 5 µL of suspension from each sample was placed on polylysine-coated glass slides and left open in the dry cabinet at 40% relative humidity and at 25 °C for 20 h. Before analysis in the SEM, the slides were coated by gold/palladium coating (20:80), at 12 mA voltage and 1 nm thickness.

2.6. Analysis of Survival Rates Following Freeze-Drying

The co-culture and mono-culture samples were generated as described above. Afterward, the samples were harvested by centrifugation at 4000× *g* for 10 min. One milliliter of each sample was washed and resuspended with DW; the suspensions were then mixed with an equal volume of skim milk (10%), to optimize the cell viability [34]. The samples were placed in a −80 °C freezer for 48 h and subsequently freeze-dried in lyophilizer (Ilshin, Hialeah, FL, USA) for 24 h. The survival rates following the freeze-drying are expressed as the number of CFUs/mL. The cell survivability (following freeze-drying) was studied during incubation of the samples in pH 2 (with 1 M hydrochloric acid) at 37 °C, for either 1 h or 3 h, using the CFU counts.

2.7. Analysis of Survival Rates Following Transition within In Vitro Digestion System

To analyze the survivability of the LAB during the passage in the gastrointestinal tract, the freeze-dried samples were resuspended in 5 mL of DW. Afterward, the samples were monitored for four hours through an in vitro digestion system using the method described previously [35].

2.8. Determining the Effect of Conditioning Supernatant (CSN) on S. aureus Biofilm Formation

Cells of *B. subtilis* and *Lactobacillus plantarum* were grown either in monoculture or co-culture using MMRS medium as described above for 24 h; *B. subtilis* cells were also grown in LB medium. Cultures were then centrifuged at 13,000 rpm for 5 min, and the supernatants were sterilized by passing through a 0.45-µm filter (Merck, Rockland, MA, USA). For analysis of the CSN effect on biofilm formation by pathogenic bacteria, the *S. aureus* starter culture was generated through its growth into 10 mL of TSB medium overnight at 37 °C and 150 rpm. The, *S. aureus* biofilm was grown into 1 mL TSB medium (within a 24-well culture plate) supplemented by the CSN (10% v/v) harvested from the above probiotic cultures. The TSB medium without supernatant was used as control. The plate was incubated for 24 h at 37 °C.

2.9. Biofilm Quantitation Assay

Crystal violet staining was performed similarly as described previously [34]. Briefly, following 24 h of incubation at 37 °C, unattached cells were removed by washing the well plates two times using DW. Then, 1% crystal violet (CV) solution was added to the wells. Following 2 min of incubation, the excess CV was removed by washing with DW. Afterward, the fixed CV was released by 33% acetic acid washing. Then, 100 µL of each sample was transferred to a new well plate for the absorbance detection at 570 nm. To confirm the CV results, the CFU quantitation was performed for surface-attached cells. Following 24 h of incubation at 37 °C, unattached cells were removed by washing the well plate two times using DW. Then, 1 mL of DW was added to each well, and the cells attached to the surface were scratched out using sterilized swab; the bacterial suspensions were serially diluted and underwent CFU counting on LB agar plates.

2.10. Effect of Cell-Free Culture Supernatant on S. aureus Growth

The, *S. aureus* starter culture was prepared into 10 mL of TSB medium overnight at 37 °C and 150 rpm. Then, 150 μL of the generated bacterial suspension was introduced into 15 mL of fresh TSB medium. For the antibacterial test, the *S. aureus* suspension was supplemented by the cell-free supernatant prepared as described above. The samples were incubated for 8 h at 37 °C at 150 rpm and subjected to OD_{600} measurements every 1 h.

2.11. Confocal Laser Scan Microscopy (CLSM) Analysis

The, *S. aureus* biofilm was grown in a confocal microscopy dish (glass-bottom dish) (Bar-Naor, Petach-Tikwa, Israel) with or without supplementation of the supernatant at the same conditions as described above. After 24 h of incubation at 37 °C, the surface-unattached cells were removed by washing the dish two times using DW. Next, the biofilm cells were stained using FilmTracer LIVE/DEAD Biofilm Viability Kit (Molecular Probes, Eugene, OR, USA) and incubated for 30 min in room temperature without exposure to light. Then, the stain was washed away and analyzed by confocal laser microscopy (CLSM) (Olympus, Hamburg, Germany). Fluorescence emission from the stained samples was measured with an SP8 CLSM (Leica) equipped with 488- and 552-nm lasers.

2.12. Real-Time PCR

To further analyze the potential antimicrobial effect of *B. subtilis* grown in MMRS medium, we tested the expression of genes that could be affected during mitigating biofilm formation by *S. aureus*. A starter culture of *B. subtilis* was prepared during overnight growth at 30 °C, 150 rpm, in LB medium. For generating the antimicrobial substance producing a suspension of *B. subtilis*, a portion of starter culture was introduced (by 1:100 ratio) into either MMRS or LB medium (as a control medium for a low antimicrobial substance production). The samples were incubated for 6 h at 37 °C, 50 rpm. Next, 2 mL from each sample was collected and centrifuged at 5000× *g* for 10 min. The RNA was harvested using the RNAeasy kit (QIAGEN, Hilden, Germany) following the manufacturer's protocol. The RNA concentration was measured by means of a Nanodrop 2000 spectrophotometer (ThermoFisher Scientific, Waltham, MA, USA). A complementary DNA (cDNA) was synthesized from 1 μg of RNA in a reverse transcription reaction using a qScript cDNA Synthesis Kit (Quantabio, Beverly, MA, USA) according to the manufacturer's instructions. All cDNA samples were stored at −20 °C. The RT-PCR reactions (final volume = 20.0 μL) consisted of 2 μL of cDNA template, 10 μL of fast SYBR green master mix, 1 μL of suspension of each primer, and 7 μL of RNase free water. Forward and reverse PCR primers (Table S2, Supplementary Materials) were designed using the Primer express software and were synthesized by Hylabs (Rehovot, Israel). DNA was amplified with the Applied Biosystems StepOne™ Real-Time PCR System (Life technologies, Foster, CA, USA) under the following PCR conditions: initial denaturation for 2 min at 95 °C and subsequent 40 PCR cycles (95 °C for 3 s, 60 °C for 30 s, and 95 °C for 15 s). The RNA samples without reverse transcriptase were used as negative control, to confirm that there was no DNA contamination in the RNA samples. The expression levels of the tested genes (*fenA, srfA*) were normalized using the 16S ribosomal RNA (rRNA) and *rpoB* genes as the endogenous controls (Table S2, Supplementary Materials).

2.13. Statistical Analysis

The results were subjected to either Student's *t*-test or one-way analysis of variance at a significance level of $p < 0.05$, to compare the control and tested samples. The results are based on three biological repeats performed in duplicates.

3. Results

3.1. Formation of Mutual Probiotic Biofilm of B. subtilis with Lactic Acid Bacteria (LAB)

The starting point of this investigation was generating the dual-species biofilms for the different LAB strains together with robust ECM-producing bacterium *B. subtilis*. Thus, the bacterial cells were incubated in the biofilm-promoting MMRS medium, which promotes increased biofilm formation by *B. subtilis* through the KinD-Spo0A pathway [33]. To visualize the mutual biofilms, a transcriptional fusion of the *tapA* promoter to the *cfp* gene (encoding cyan fluorescent protein) was used [35]. The observed upregulation in the CFP expression, during the mutual growth of *B. subtilis* with three different species of the probiotic LAB, indicates that the *tapA* operon was activated and there that there was notable matrix production by *B. subtilis* (Figure 1). This finding was quite noticeable following a comparison of morphological changes that occurred during LAB growth in the presence of *B. subtilis* (Figure S1, Supplementary Materials). In this regard, the LAB cells could not form any biofilm bundles during their growth as a mono-species culture (Figure S1, Supplementary Materials), whereas a notable incorporation of those cells was observed into biofilm bundles produces by *B. subtilis* cells (Figure 1).

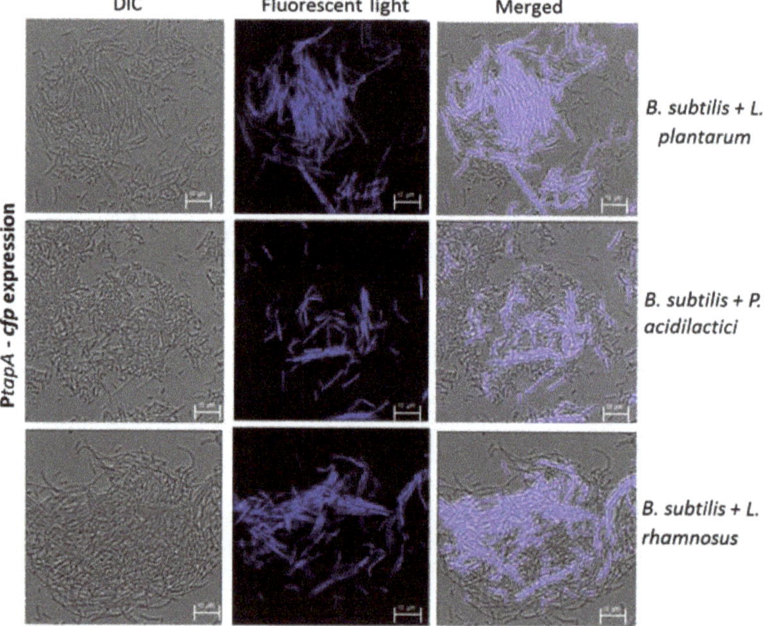

Figure 1. Formation of dual-species biofilm bundles of *B. subtilis* with LAB. Bacterial biofilms were generated during co-culture growth of *B. subtilis* cells with either *Lactobacillus plantarum*, *Pediococcus acidilactici*, or *L. rhamnosus* cells in modified MRS (MMRS) medium at 37 °C for 8 h. The biofilm samples were prepared as described in Section 2 and analyzed using a confocal laser scanning microscope (CSLM, Leica, Germany). *B. subtilis* cells express CFP under the control of the *tapA* operon, which is responsible for the matrix production. LAB cells are not stained. Scale bar = 10 µm.

To confirm that there are no antagonistic interactions between the LAB and *B. subtilis* cells, the bacterial growth was analyzed in this mutual growth system. Consequently, there was no significant inhibition in either of the bacterial species following their mutual growth (Figure 2), meaning that the LAB and *B. subtilis* cells can grow together without interference through generating the mutual probiotic biofilm. In addition, the growth in dual-species biofilm did not change the medium acidification rate by the LAB cells (Figure S2, Supplementary Materials).

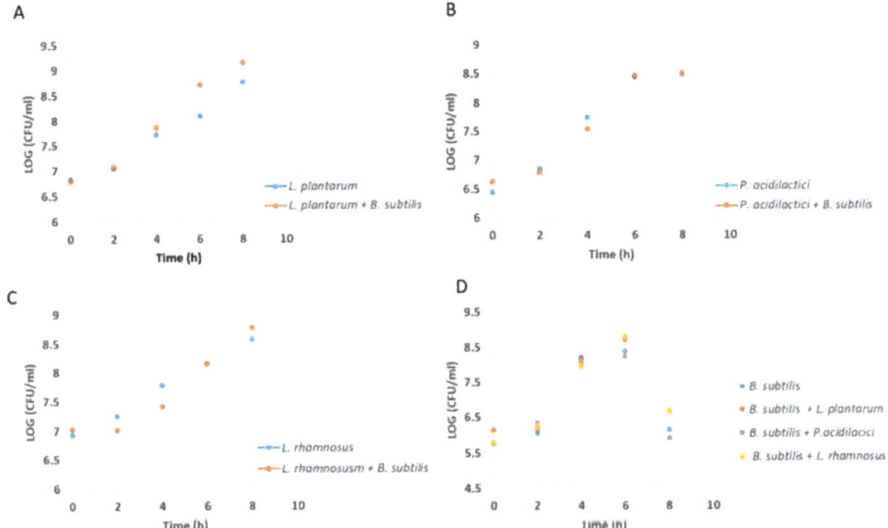

Figure 2. Co-culture growth of *B. subtilis* with the LAB bacteria does not influence growth rate. Growth curve analysis of LAB cells in presence or absence of *B. subtilis*. The blue line represents the growth rate of a single-species culture of LAB, whereas the orange line represents the growth rate of (**A**) *L. plantarum*, (**B**) *P. acidilactici*, and (**C**) *L. rhamnosus* in co-culture with *B. subtilis*. (**D**) Growth curve of *B. subtilis* cells in the presence of LAB species compared to single culture. The cells were analyzed during growth for 8 h in 37 °C, 150 rpm.

3.2. Growth in Mutual Biofilm Increases the Survivability of the LAB during Desiccation

It was hypothesized that the growth in the mutual biofilm system could provide a relative protection of LAB during the desiccation process, which might indicate about the relative improvement in survivability of the bio-coated cells through industrial processing and storage conditions. Therefore, the LAB cells grown in the mutual biofilm were exposed to desiccation conditions for either 20 or 40 h. The LAB cells grown in mono-species culture were used as a control sample. It was found that *L. plantarum* cells grown in mutual biofilm showed increased survival (relatively to mono-species culture) of around 1.12 log·CFU/mL and 1.52 log·CFU/mL, following 20 and 40 h of desiccation, respectively. Surprisingly, *L. rhamnosus* cells grown in the mutual biofilm demonstrated an even more significant increase in survivability, of around 3.12 log·CFU/mL, during 20 h of desiccation. Even more profoundly, the bio-coated cells of *L. rhamnosus* demonstrated around a five-log increase in their survivability after 40 h of desiccation. Concerning the cells of *P. acidilactici*, the bio-coated cells showed a relatively moderate increased survival following 20 and 40 h of desiccation of around 0.71 log·CFU/mL and 2.09 log·CFU/mL, respectively (Figure 3).

To reinforce our assumption about the ECM protection of the LAB cells, the dual- and mono-species biofilms were visualized using SEM imaging, after desiccation treatment. It was found that *L. plantarum* cells grown in the presence of *B. subtilis* were surrounded with the coating substance(s), which could be interpreted as the biofilm matrix (Figure 4).

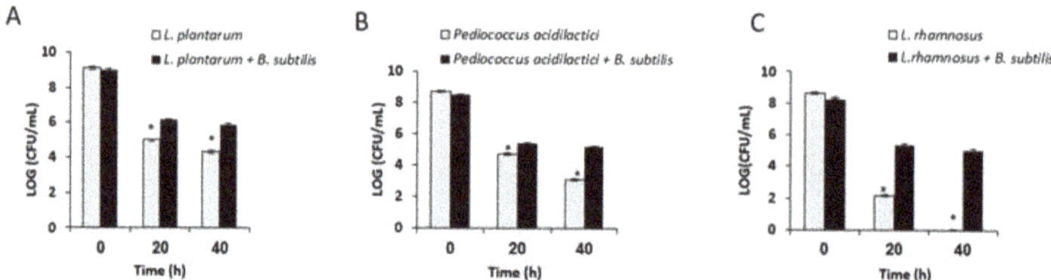

Figure 3. Growth in dual-species biofilm increases survivability of LAB during exposure to drying process. Mono and dual-species cultures of *B. subtilis* and LAB cells were generated in MMRS medium during bacterial growth for 8 h in 37 °C, 50 rpm. Survival rates of the LAB cells (grown in presence or absence of biofilm forming *B. subtilis*) were determined based on CFU counts following desiccation conditions (40% relative humidity (RH)) for 20–40 h. Error bars represent standard deviation (SD). * $p < 0.05$ comparison of the control and tested samples. (**A**) *L. plantarum* survival rates, (**B**) *P. acidilactici* survival rates, and (**C**) *L. rhamnosus* survival rates.

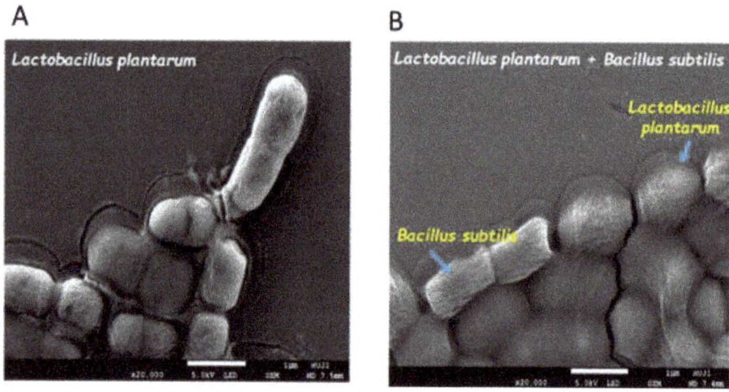

Figure 4. Bio-coating of *L. plantarum* cells by extracellular polymeric substance(s) produced by *B. subtilis*. Mono and dual-species cultures of *B. subtilis* and *L. plantarum* cells were generated in MMRS medium during bacterial growth for 8 h in 37 °C, 50 rpm. (**A**) SEM images of the mono-species culture of *L. plantarum* and (**B**) dual-species culture of *L. plantarum* with *B. subtilis* following desiccation conditions (40% RH) for 20 h. Images were taken at a magnification of 20,000× with a JEOL, JSM 7800F, Japan.

3.3. Growth in Mutual Biofilm Increases the Survivability of the LAB during Acid Stress Following Freeze-Drying

It was further investigated whether the mutual biofilm could also protect the cells of LAB during freeze-drying, which is considered as the most common technique for drying and storage of probiotic bacteria for a long time [10,36]. Since, after consumption, the LAB cells are usually exposed to additional stress (acidic stress during the passage in the gastrointestinal tract), the survivability of the bio-coated cells during their exposure to acid stress was tested, following freeze-drying. To mimic the acid stress conditions, the bio-coated cells were freeze-lyophilized and exposed to pH 2. It was found that the bio-coated *L. plantrum* cells showed increased survivability of around 0.45 log·CFU/mL, compared to the control, following freeze-drying (Figure 5). In the case of the bio-coated *L. rhamnosus* cells, an increase of about 0.49 log·CFU/mL was observed, while, in the case of *P. acidilactici*, there was no significant change in the number of viable cells following freeze-drying (Figure 5). Concerning acid stress tolerance, the freeze-dried LAB cells were exposed to low pH (pH 2) for 1–3 h. As shown in Figure 5, the encapsulated LAB consistently demonstrated increased survival (up to around a one-log increase) following their exposure to this stress, especially after three hours of exposure to the low pH.

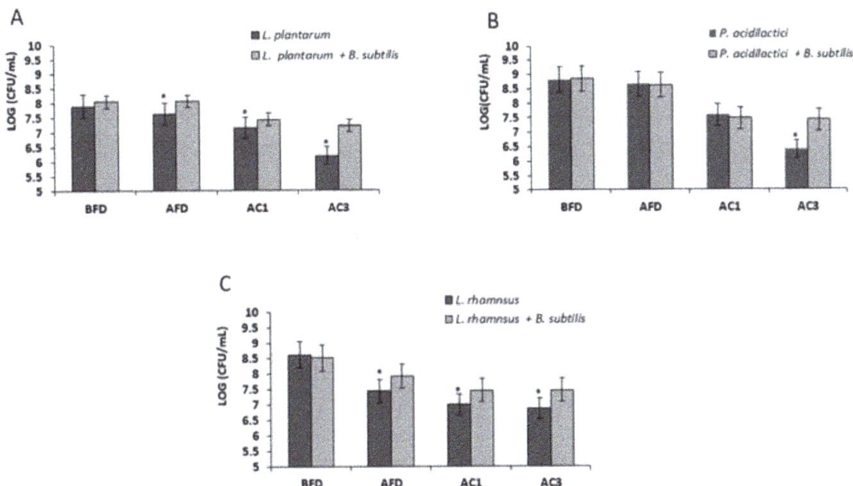

Figure 5. Bio-coating increases the survivability of the LAB during acid stress following freeze-drying. Mono- and dual-species cultures of *B. subtilis* and the LAB cells were generated in MMRS medium during bacterial growth for 8 h in 37 °C, 50 rpm. Survival rates of LAB cells were determined based on CFU counts following freeze-drying and an exposure to low pH (HCl 1 M, pH = 2) for 1–3 h. * $p < 0.05$ for comparison of the control and tested samples. Error bars represent standard deviation (SD). BFD—before freeze-drying; AFD—after freeze-drying; AC1—freeze-dried and exposed for 1 h to acid conditions; AC3—freeze-dried and exposed for 3 h to acid conditions. The survival rates are shown for (**A**) *L. plantarum*, (**B**) *P. acidilactici*, and (**C**) *L. rhamnosus*.

3.4. Bio-Coating Retains the LAB Survivability during In Vitro Gastrointestinal Digestion Following Freeze-Drying

It was further tested whether bio-coating could also provide protection of the LAB cells during passage through an in vitro GIT system. This unique system included two phases: a gastric phase characterized by low pH and stomach proteolytic enzymes and, the intestinal phase, characterized by neutral pH, intestinal proteolytic enzymes, and bile salts. The bio-coated LAB cells were freeze-lyophilized and tested through this system. It was found that the bio-coated cells of *L. plantarum* and *L. rhamnosus* showed increased survivability of around 0.5 log following freeze-drying, which remained at the augmented level during the transition of those cells through the GIT system (Figure 6). Nonetheless, in the case of *P. acidilactici*, there was no significant increase in the survivability of the cells in the tested conditions.

3.5. Antagonistic Effect of Probiotic Cells against the Biofilm-Forming, S. aureus

Next, it was hypothesized that the probiotic cells (from the mutual biofilm) could antagonize pathogenic bacteria, for instance, *S. aureus*, which is known as a robust biofilm-forming bacterium, especially a submerged type of biofilm. It was consequently found that the conditioning supernatant (CSN) obtained during the growth of the probiotic cells strongly inhibited biofilm formation by *S. aureus* (Figure 7A). This result indicates that there might be an induction in producing an antimicrobial substance(s) during the generation of the mutual biofilm. Interestingly, it seems that a major impact of this inhibitory effect was related to *B. subtilis* cells, although there was a modest contribution by the cells of *L. plantarum*. A further quantitation of the surface-adhered cells confirmed a potent inhibitory effect of the CSN against biofilm formation by pathogenic *S. aureus* (Figure 7B). Moreover, the microscopic visualization of the augmented biofilm phenotypes confirmed once again the anti-staphylococcal properties of the CSN (Figure 7C). Importantly, it was further confirmed that the CSN did not cause growth inhibition of *S. aureus* (Figure 7D), which indicates the biofilm-specific mode of action of CNS against this pathogenic bacterium.

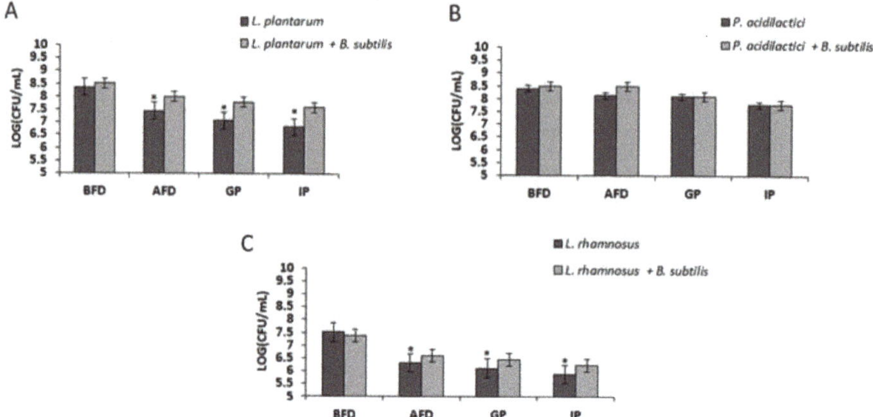

Figure 6. Bio-coating maintains LAB survivability during gastrointestinal digestion in vitro following freeze-drying. Mono- and dual-species cultures of *B. subtilis* and LAB cells were generated in MMRS medium during bacterial growth for 8 h in 37 °C, 50 rpm. Survival rates of LAB cells were determined based on CFU counts following freeze-drying and during gastro-intestinal digestion in vitro. * $p < 0.05$ for comparison of the control and tested samples. Error bars represent standard deviation (SD). BFD—before freeze-drying; AFD—after freeze-drying; GP—gastric phase; IP—intestinal phase. The survival rates are shown for (**A**) *L. plantarum*, (**B**) *P. acidilactici*, and (**C**) *L. rhamnosus*.

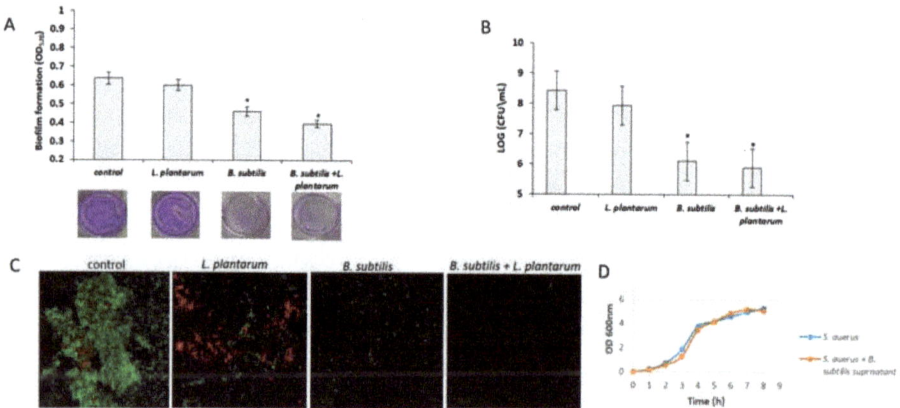

Figure 7. Conditioning supernatant (CNS) of the *B. subtilis* culture inhibits biofilm formation by *Staphylococcus aureus*. *S. aureus* biofilm formation was determined after growth in TSB medium with 10% CNS at 37 °C for 24 h. (**A**) *S. aureus* biofilm formation quantification by crystal violet method. (**B**) Quantification of live bacteria attached to surface using CFU method. * $p < 0.05$ for comparison of the control and tested samples. Error bars represent standard deviation (SD). (**C**) Confocal laser microscopy (CLSM) images of *S. aureus* biofilms formed onto polystyrene surfaces containing 10% CNS. Live cells (SYTO-9, green) and dead cells (propidium iodide, red). Scale bar = 50 μm. (**D**) *S. aureus* growth curve in TSB medium supplemented by the 10% CNS generated from the *B. subtilis* growth medium.

It was further hypothesized that the growth media would affect the ability of the CSN to inhibit *S. aureus* biofilm formation. Therefore, the antibiofilm activity of the CNS produced by *B. subtilis* cells in the MMRS medium was compared with that produced in LB medium. Apparently, growth of the *Bacillus* cells in the MMRS induced the antibiofilm effect of the CSN (Figure 8). Thus, a significantly higher inhibition on *S. aureus* biofilm formation was found by the CSN from the MMRS medium compared to that produced in the LB medium. Accordingly, the CSN from MMRS medium showed around a three-log reduction in the *S. aureus* adherence onto the surface compared to that from LB (Figure 8A). The inhibitory effect of the CSN was further confirmed microscopically by testing a submerged biofilm of *S. aureus* cells using live–dead staining (Figure 8B).

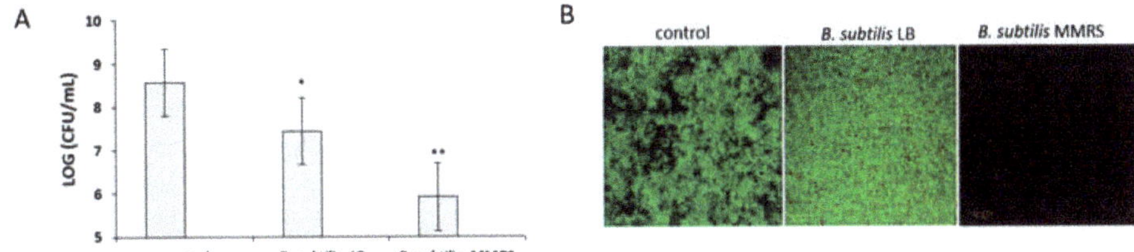

Figure 8. Type of growth medium governs the inhibitory effect on the *S. aureus* biofilm formation. (**A**) quantification of the *S. aureus* cells attached to the surface using CFU method, and (**B**) the CLSM imaging of the *S. aureus* biofilm formation in the presence or absence of the CSN, following growth in TSB medium with 10% supernatant at 37 °C for 24 h. * $p < 0.05$ for comparison of the CSN from LB to control; ** $p < 0.05$ for comparison of CSN from MMRS vs. LB. Error bars represent standard deviation (SD). Live cells are stained green (SYTO-9) and dead cells are stained red (propidium iodide). Scale bar = 50 µm.

According to recent findings, *B. subtilis* could affect *S. aureus* biofilm formation via signaling interference [4]. It was, thus, hypothesized that the production of either fengycins [4] or surfactin [37] could explain the antibiofilm activity of the CSN produced by *B. subtilis*. Both factors are produced by *B. subtilis*, and they could affect the *S. aureus* cells through interfering with inter- or intra-cellular signaling. It was subsequently found that expression of the genes encoding for these factors by *B. subtilis* was notably upregulated in the MMRS compered to LB medium (three- and two-fold induction in the expressions of *fenA* and *srfA*, respectively) (Figure 9). This result suggests the involvement of the regulatory pathways associated with those genes in the observed antibiofilm phenotype.

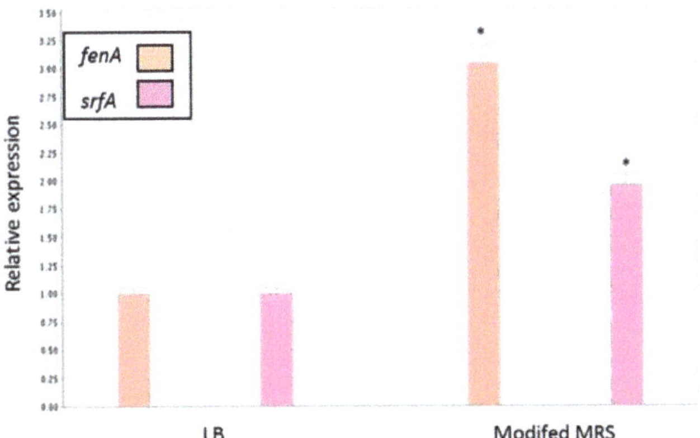

Figure 9. Relative expression of *B. subtilis* genes related to antagonistic activity during growth in the different media. The real-time (RT)-PCR analysis was performed for quantitation of *fenA* and *srfA* gene expression in *B. subtilis* cells grown in either LB or MMRS medium as described in the Methods. * $p < 0.05$ compared to control. Error bars represent standard deviation (SD).

4. Discussion

The importance of healthy commensal microbiota for the mammalian host is evident; thus, there is widespread use of probiotics for preventing and treating various health problems in humans, as well as in animals. Nonetheless, maximizing the survivability of probiotic cells during their formulation, as well delivery, remains a significant challenge. Probiotic bacteria are supposed to go through a long route starting with processing, through shelf life and the passage through the GIT, which includes dealing with extreme conditions [38]. Since these processes may affect cell survivability, an effective way(s) of delivering probiotic bacteria to the mammalian host would be highly useful.

It is now well established that biofilm formation represents one of the most favorable microbial lifestyles within often challenging natural environments [39]. The biofilm provides bacterial cells protection against challenging environmental conditions such as changes in shear forces, extreme temperatures, desiccation, extreme pH, and antimicrobial agents [40–42]. It was, therefore, proposed to generate a protective bio-coating system for probiotic cells for a possibility of a more efficient delivery to the mammalian host. The most appropriate candidate for this mission appeared to be the robust biofilm-forming *B. subtilis*, since it naturally colonizes the mammalian gut and is considered to be harmless to mammals including humans [19,20]. It was confirmed that there are no antagonistic interactions during the generation of this complex multispecies system; thus, no antagonism was observed between *B. subtilis* cells and LAB, or during the formation of symbiotic biofilm bundles through inducing the expression of *tapA* operon (involved in the matrix production) by *B. subtilis*. It should be emphasized that the tested LAB strains belong to different genera with a different origin. Nonetheless, it was possible to generate cooperating and protected multispecies biofilms, which indicates the feasibility of using this bio-coating system for a wide range of probiotic species.

As suggested throughout the study, the generated bio-coating system increased the LAB survivability during drying processes, which points to the feasibility of using this system for processing probiotic cells for further food or biotechnological applications. The drying process is commonly used as a means for storage and distribution, which lowers the expense and inconvenience of using a cool chain. Although water is essential for bacteria living and the drying processes damage cell structure and viability, the long-time preservation and retention of viability during storage is often enhanced by lowering the water activity [34]. The robust biofilm matrix, produced by *B. subtilis*, contains polysaccharides (PS) that presumably have an important role in protecting bacteria during drying processes. The PS could provide bacteria a hydrated microenvironment. Thus, through the drying process, the PS layers may serve as a barrier on the cell surface and prevent water removal [18,43]. In the case of *L. rhamnosus*, which could not survive the 40 h of desiccation, the bio-coating process enabled a very significant increase in survivability during the desiccation process. This finding highly suggests the possibility of protecting desiccation-sensitive probiotic cells using this bio-coating system.

Freeze-drying is the most common drying method for long-term preservation of microorganisms, in the microbiological industry, thanks to optimal protection of cell viability [36]. Usually, before the freeze-drying process, protective agents like skim milk, sucrose, or other sugar types are added to the drying medium to prevent cell damage during the drying process and storage of freeze-dried cells [44]. Some studies showed that biofilm PS can also be used as a protective agent [43,45], which is in agreement with the findings of current study. We observed higher survivability during freeze-drying for *L. plantarum* and *L. rhamnosus* cells following their growth through the bio-coating system. However, we did not observe a significant increase in survival rates for *P. acidilactici* following freeze-drying. One of the possible explanations for this result could be related to the possible resistance of this environmental isolate to desiccation stress due to its adaptation to the udder environment (from where it was isolated).

In addition to the protective capability, *B. subtilis* demonstrated potent antimicrobial activity against pathogenic *S. aureus*. This result was not surprising since *B. subtilis* was recently explored for its probiotic functionality on many levels. It was shown that *B. subtilis* could stimulate an immune response in humans, as well as maintain a favorable balanced microbiota, and decrease infection and diarrhea via the synthesis of antimicrobial agents [46]. Production of antimicrobial agents is one of the antagonistic properties of probiotic bacteria, and indeed *B. subtilis* produces a wide diversity of substances, which influence a broad spectrum of pathogens via different mechanisms [27]. Several studies suggested either the growth inhibition or depression of *Staphylococcal* virulence by the antagonistic activity of *B. subtilis* [4,47]. In agreement with the literature, the current study showed that most of the antimicrobial activity of the multispecies biofilm system was due to substances produced by *B. subtilis*. Importantly, this activity was specific to the mitigation of biofilm formation by *S. aureus* rather than inhibition of its growth. Notably, the relatively modest effect of the CSN derived from *L. plantarum* on biofilm formation by *S. aureus* cells might still have an important role in mitigating this problematic pathogen. The synergistic activity of substances produced by different probiotic species might provide a further antimicrobial effect against such persistent pathogens.

Findings of this study further indicated that the inhibitory effect of the CSN is associated with the production of secondary metabolites, for instance, lipopeptides by *B. subtilis*. In this regard, the inhibitory effect could be related to their chemical structure [30] or to their ability to inhibit quorum sensing [4]. Lipopeptides produced by *B. subtilis* function firstly as quorum-sensing interrupters (fengycins), by inhibiting the quorum-sensing regulatory system [4], and surfactin, by regulating the autoinducer-2 (AI-2) activity [38]. Thus, these lipopeptides could inhibit quorum sensing via a different mechanism. The other antimicrobial mode of action of lipopeptides is related to the similarity of their chemical structure to surface-active agents, which might impair the ability of cells to attach to the surface and form a biofilm structure [30]. According to the results presented in this study, it appears that the growth of *B. subtilis* in the MMRS medium triggers the production of antimicrobial lipopeptides. This finding is indeed conceivable since the MMRS medium notably induces biofilm formation in *B. subtilis* [33], which is highly related to the production of antimicrobial lipopeptides [32].

Taken together, the data shown in this study suggest of the robust probiotic functionality of *B. subtilis* (i) in protecting the probiotic LAB during their exposure to unfavorable environmental conditions, such as desiccation and acid stresses, and (ii) strong anti-biofilm activity against pathogenic bacteria such as *S. aureus*.

Supplementary Materials: The following are available online at http://www.mdpi.com/2076-2607/7/10/407/s1: Figure S1: The CSLM images of LAB cells following their monoculture growth; Figure S2: Growth in dual-species biofilm does not change medium acidification rate by LAB cells; Table S1: Bacterial strains used in this study; Table S2: Primers used for RT-PCR analyses.

Author Contributions: Conceptualization was originated by M.S. and developed together with H.K.; developing a methodology during the investigation and the data analysis accomplished by H.K. and M.S.; the experiments were performed by H.K; the writing, review & editing of the manuscript accomplished by H.K and M.S.; the supervision and funding acquisition were done by M.S. All authors approved the final version of the manuscript.

Funding: This work was partially supported by the Nitzan Grant No. 4210376 of the Chief Scientist of the Ministry of Agriculture and Rural Development (Israel). Hadar Kimelman is a recipient of an Excellence Scholarship for M.Sc. students granted by the Agricultural Research Organization, as well as by The Hebrew University in Jerusalem.

Acknowledgments: This study forms part of Hadar Kimelman's M.Sc. project. We would like to acknowledge Doron Steinberg (from the Hebrew University of Jerusalem) for helpful discussions. We also would like to thank Yulia Kroupitski, Yigal Achmon, Carmel Hutchings, and Bat-Chen Cohen (from the Shemesh lab, ARO), Shamay Yaacoby (from the Department of Ruminant Science, ARO), and Shlomo Blum, Marcelo Fleker, and Oleg Krifucks (from the National Mastitis Reference Center, Kimron Veterinary Institute) for their help in bacterial isolation and characterization. We are also grateful to Ram Reifen and Zipi Berkovich (from The Hebrew University of Jerusalem) for their assistance performing experiments in the in vitro GIT system. We also acknowledge members of the Shemesh lab for helpful discussions and technical assistance. We also wish to thank Eduard Belausov (from the ARO) and Dr. Einat Zelinger (from the Hebrew University) for excellent technical assistance with the microscopy analyses.

Conflicts of Interest: The authors declare no conflicts of interest.

References

1. Reid, G.; Jass, J.; Sebulsky, M.T.; McCormick, J.K. Potential uses of probiotics in clinical practice. *Clin. Microbiol. Rev.* **2003**, *16*, 658–672. [CrossRef] [PubMed]
2. Holzapfel, W.H.; Haberer, P.; Geisen, R.; Björkroth, J.; Schillinger, U. Taxonomy and important features of probiotic microorganisms in food and nutrition. *Am. J. Clin. Nutr.* **2001**, *73*, 365S–373S. [CrossRef]
3. Anukam, K.C.; Reid, G. Probiotics: 100 years (1907–2007) after Elie Metchnikoff's Observation. *Commun. Curr. Res. Educ. Top. Trends Appl. Microbiol.* **2007**, *1*, 466–474.
4. Piewngam, P.; Zheng, Y.; Nguyen, T.H.; Dickey, S.W.; Joo, H.S.; Villaruz, A.E.; Chiou, J.; Glose, K.A.; Fisher, E.L.; Hunt, R.L.; et al. Pathogen elimination by probiotic Bacillus via signalling interference. *Nature* **2018**, *562*, 532.
5. Cotter, P.D.; Hill, C.; Ross, R.P. Bacteriocins: Developing innate immunity for food. *Nat. Rev. Microbiol.* **2005**, *3*, 777–788. [CrossRef]
6. Servin, A.L. Antagonistic activities of lactobacilli and bifidobacteria against microbial pathogens. *FEMS Microbiol. Rev.* **2004**, *28*, 405–440. [CrossRef] [PubMed]
7. Aw, W.; Fukuda, S. Protective effects of bifidobacteria against enteropathogens. *Microb. Biotechnol.* **2019**. [CrossRef]
8. Fukuda, S.; Toh, H.; Hase, K.; Oshima, K.; Nakanishi, Y.; Yoshimura, K.; Tobe, T.; Clarke, J.M.; Topping, D.L.; Suzuki, T.; et al. Bifidobacteria can protect from enteropathogenic infection through production of acetate. *Nature* **2011**, *469*, 543–547. [CrossRef]
9. Iaconelli, C.; Lemetais, G.; Kechaou, N.; Chain, F.; Bermúdez-Humarán, L.G.; Langella, P.; Gervais, P.; Beney, L. Drying process strongly affects probiotics viability and functionalities. *J. Biotechnol.* **2015**, *214*, 17–26. [CrossRef]
10. Marcial-Coba, M.S.; Knochel, S.; Nielsen, D.S. Low-moisture food matrices as probiotic carriers. *FEMS Microbiol. Lett.* **2019**, *366*. [CrossRef]
11. Sanchez, B.; Champomier-Vergès, M.C.; Collado Mdel, C.; Anglade, P.; Baraige, F.; Sanz, Y.; de los Reyes-Gavilán, C.G.; Margolles, A.; Zagorec, M. Low-pH adaptation and the acid tolerance response of Bifidobacterium longum biotype longum. *Appl. Environ. Microbiol.* **2007**, *73*, 6450–6459. [CrossRef] [PubMed]
12. Bron, P.A.; Marco, M.; Hoffer, S.M.; Van Mullekom, E.; De Vos, W.M.; Kleerebezem, M. Genetic characterization of the bile salt response in Lactobacillus plantarum and analysis of responsive promoters in vitro and in situ in the gastrointestinal tract. *J. Bacteriol.* **2004**, *186*, 7829–7835. [CrossRef] [PubMed]
13. Justice, S.S.; Harrison, A.; Becknell, B.; Mason, K.M. Bacterial differentiation, development, and disease: Mechanisms for survival. *FEMS Microbiol. Lett.* **2014**, *360*, 1–8. [CrossRef] [PubMed]
14. Wingender, J.; Neu, T.R.; Flemming, H.C. What are Bacterial Extracellular Polymeric Substances? In *Microbial Extracellular Polymeric Substances*; Springer: Berlin/Heidelberg, Germany, 1999; pp. 1–19.
15. Flemming, H.C.; Wingender, J. The biofilm matrix. *Nat. Rev. Microbiol.* **2010**, *8*, 623–633. [CrossRef] [PubMed]
16. Webb, J.S.; Givskov, M.; Kjelleberg, S. Bacterial biofilms: Prokaryotic adventures in multicellularity. *Curr. Opin. Microbiol.* **2003**, *6*, 578–585. [CrossRef] [PubMed]
17. Roberson, E.B.; Firestone, M.K. Relationship between Desiccation and Exopolysaccharide Production in a Soil *Pseudomonas* sp. *Appl. Environ. Microbiol.* **1992**, *58*, 1284–1291. [PubMed]
18. Limoli, D.H.; Jones, C.J.; Wozniak, D.J. Bacterial Extracellular Polysaccharides in Biofilm Formation and Function. *Microbiol. Spectr.* **2015**, *3*. [CrossRef] [PubMed]
19. Ilinskaya, O.N.; Ulyanova, V.V.; Yarullina, D.R.; Gataullin, I.G. Secretome of Intestinal Bacilli: A Natural Guard against Pathologies. *Front. Microbiol.* **2017**, *8*, 1666. [CrossRef] [PubMed]
20. Batista, M.T.; Souza, R.D.; Paccez, J.D.; Luiz, W.B.; Ferreira, E.L.; Cavalcante, R.C.; Ferreira, R.C.; Ferreira, L.C. Gut Adhesive Bacillus subtilis Spores as a Platform for Mucosal Delivery of Antigens. *Infect. Immun.* **2014**, *82*, 1414–1423. [CrossRef]
21. Shemesh, M.; Chai, Y. A combination of glycerol and manganese promotes biofilm formation in Bacillus subtilis via histidine kinase KinD signaling. *J. Bacteriol.* **2013**, *195*, 2747–2754. [CrossRef]
22. Pasvolsky, R.; Zakin, V.; Ostrova, I.; Shemesh, M. Butyric acid released during milk lipolysis triggers biofilm formation of Bacillus species. *Int. J. Food Microbiol.* **2014**, *181*, 19–27. [CrossRef] [PubMed]
23. Vlamakis, H.; Chai, Y.; Beauregard, P.; Losick, R.; Kolter, R. Sticking together: building a biofilm the Bacillus subtilis way. *Nat. Rev. Microbiol.* **2013**, *11*, 157–168. [CrossRef] [PubMed]

24. Lefevre, M.; Racedo, S.M.; Ripert, G.; Housez, B.; Cazaubiel, M.; Maudet, C.; Jüsten, P.; Marteau, P.; Urdaci, M.C. Probiotic strain Bacillus subtilis CU1 stimulates immune system of elderly during common infectious disease period: a randomized, double-blind placebo-controlled study. *Immun. Ageing.* **2015**, *12*, 24. [CrossRef] [PubMed]
25. Lefevre, M.; Racedo, S.M.; Denayrolles, M.; Ripert, G.; Desfougères, T.; Lobach, A.R.; Simon, R.; Pélerin, F.; Jüsten, P.; Urdaci, M.C. Safety assessment of Bacillus subtilis CU1 for use as a probiotic in humans. *Regul. Toxicol. Pharmacol.* **2017**, *83*, 54–65. [CrossRef] [PubMed]
26. Hosoi, T.; Ametani, A.; Kiuchi, K.; Kaminogawa, S. Improved growth and viability of lactobacilli in the presence of Bacillus subtilis (natto), catalase, or subtilisin. *Can. J. Microbiol.* **2000**, *46*, 892–897. [CrossRef]
27. Urdaci, M.C.; Pinchuk, I. Antimicrobial activity of Bacillus probiotics. In *Bacterial Spore Formers–Probiotics and Emerging Applications*; Horizon Bioscience: Norfolk, UK, 2004; pp. 171–182.
28. Khochamit, N.; Siripornadulsil, S.; Sukon, P.; Siripornadulsil, W. Antibacterial activity and genotypic-phenotypic characteristics of bacteriocin-producing Bacillus subtilis KKU213: Potential as a probiotic strain. *Microbiol. Res.* **2015**, *170*, 36–50. [CrossRef]
29. Zhao, H.; Yan, L.; Xu, X.; Jiang, C.; Shi, J.; Zhang, Y.; Liu, L.; Lei, S.; Shao, D.; Huang, Q. Potential of Bacillus subtilis lipopeptides in anti-cancer I: induction of apoptosis and paraptosis and inhibition of autophagy in K562 cells. *AMB Express* **2018**, *8*, 78. [CrossRef]
30. Zhao, H.; Shao, D.; Jiang, C.; Shi, J.; Li, Q.; Huang, Q.; Rajoka, M.S.R.; Yang, H.; Jin, M. Biological activity of lipopeptides from Bacillus. *Appl. Microbiol. Biotechnol.* **2017**, *101*, 5951–5960. [CrossRef]
31. Zeriouh, H.; de Vicente, A.; Pérez-García, A.; Romero, D. Surfactin triggers biofilm formation of Bacillus subtilis in melon phylloplane and contributes to the biocontrol activity. *Environ. Microbiol.* **2014**, *16*, 2196–2211. [CrossRef]
32. López, D.; Fischbach, M.A.; Chu, F.; Losick, R.; Kolter, R. Structurally diverse natural products that cause potassium leakage trigger multicellularity in Bacillus subtilis. *Proc. Natl. Acad. Sci. USA* **2009**, *106*, 280–285. [CrossRef]
33. Yahav, S.; Berkovich, Z.; Ostrov, I.; Reifen, R.; Shemesh, M. Encapsulation of beneficial probiotic bacteria in extracellular matrix from biofilm-forming Bacillus subtilis. *Artif. Cells Nanomed. Biotechnol.* **2018**, *46*, 974–982. [CrossRef] [PubMed]
34. Meng, X.C.; Stanton, C.; Fitzgerald, G.F.; Daly, C.; Ross, R.P. Anhydrobiotics: The challenges of drying probiotic cultures. *Food Chem.* **2008**, *106*, 1406–1416. [CrossRef]
35. Chai, Y.; Chu, F.; Kolter, R.; Losick, R. Bistability and biofilm formation in Bacillus subtilis. *Mol. Microbiol.* **2008**, *67*, 254–263. [CrossRef] [PubMed]
36. Morgan, C.A.; Herman, N.; White, P.A.; Vesey, G. Preservation of micro-organisms by drying; a review. *J. Microbiol. Methods* **2006**, *66*, 183–193. [CrossRef] [PubMed]
37. Liu, J.; Li, W.; Zhu, X.; Zhao, H.; Lu, Y.; Zhang, C.; Lu, Z. Surfactin effectively inhibits Staphylococcus aureus adhesion and biofilm formation on surfaces. *Appl. Microbiol. Biotechnol.* **2019**, *103*, 4565–4574. [CrossRef] [PubMed]
38. Flores-Belmonta, I.A.; Palou, E.; López-Malo, A.; Jiménez-Munguía, M.T. Simple and double microencapsulation of Lactobacillus acidophilus with chitosan using spray drying. *Int. J. Food Stud.* **2015**. [CrossRef]
39. Watnick, P.; Kolter, R. Biofilm, city of microbes. *J. Bacteriol* **2000**, *182*, 2675–2679. [CrossRef] [PubMed]
40. Yin, W.; Wang, Y.; Liu, L.; He, J. Biofilms: The Microbial "Protective Clothing" in Extreme Environments. *Int. J. Mol. Sci.* **2019**, *20*, 3423. [CrossRef] [PubMed]
41. Shemesh, M.; Pasvolsky, R.; Zakin, V. External pH is a cue for the behavioral switch that determines surface motility and biofilm formation of Alicyclobacillus acidoterrestris. *J. Food Prot* **2014**, *77*, 1418–1423. [CrossRef] [PubMed]
42. Shemesh, M.; Kolter, R.; Losick, R. The Biocide Chlorine Dioxide Stimulates Biofilm Formation in Bacillus subtilis by Activation of the Histidine Kinase KinC. *J. Bacteriol.* **2010**, *192*, 6352–6356. [CrossRef] [PubMed]
43. Nguyen, H.T.; Razafindralambo, H.; Blecker, C.; N'Yapo, C.; Thonart, P.; Delvigne, F. Stochastic exposure to sub-lethal high temperature enhances exopolysaccharides (EPS) excretion and improves Bifidobacterium bifidum cell survival to freeze-drying. *Biochem. Eng. J.* **2014**, *88*, 85–94. [CrossRef]
44. Carvalho, A.S.; Silva, J.; Ho, P.; Teixeira, P.; Malcata, F.X.; Gibbs, P. Relevant factors for the preparation of freeze-dried lactic acid bacteria. *Int. Dairy J.* **2004**, *14*, 835–847. [CrossRef]

45. Nguyen, H.T.; Truong, D.H.; Kouhoundé, S.; Ly, S.; Razafindralambo, H.; Delvigne, F. Biochemical Engineering Approaches for Increasing Viability and Functionality of Probiotic Bacteria. *Int. J. Mol. Sci.* **2016**, *17*, 867. [CrossRef] [PubMed]
46. Suva, M.A.; Sureja, V.P.; Kheni, D.B. Novel insight on probiotic Bacillus subtilis: Mechanism of action and clinical applications. *J. Curr. Res. Sci. Med.* **2016**, *2*, 65. [CrossRef]
47. Gonzalez, D.J.; Haste, N.M.; Hollands, A.; Fleming, T.C.; Hamby, M.; Pogliano, K.; Dorrestein, P.C. Microbial competition between Bacillus subtilis and Staphylococcus aureus monitored by imaging mass spectrometry. *Microbiol. Sgm.* **2011**, *157*, 2485–2492. [CrossRef] [PubMed]

Article

Production of Hydrophobic Zein-Based Films Bioinspired by The Lotus Leaf Surface: Characterization and Bioactive Properties

Ângelo Luís [1,2], Fernanda Domingues [1,3] and Ana Ramos [3,4,*]

1 Centro de Investigação em Ciências da Saúde (CICS-UBI), Universidade da Beira Interior, Avenida Infante D. Henrique, 6200–506 Covilhã, Portugal

2 Laboratório de Fármaco-Toxicologia, UBIMedical, Universidade da Beira Interior, Estrada Municipal 506, 6200–284 Covilhã, Portugal

3 Departamento de Química, Faculdade de Ciências, Universidade da Beira Interior, Rua Marquês d'Ávila e Bolama, 6201–001 Covilhã, Portugal

4 Materiais Fibrosos e Tecnologias Ambientais (FibEnTech), Universidade da Beira Interior, Rua Marquês d'Ávila e Bolama, 6201–001 Covilhã, Portugal

* Correspondence: ammr@ubi.pt; Tel.: +351-275-319-700

Received: 2 August 2019; Accepted: 14 August 2019; Published: 16 August 2019

Abstract: Hydrophobic zein-based functional films incorporating licorice essential oil were successfully developed as new alternative materials for food packaging. The lotus-leaf negative template was obtained using polydimethylsiloxane (PDMS). The complex surface patterns of the lotus leaves were transferred onto the surface of the zein-based films with high fidelity (positive replica), which validates the proposed proof-of-concept. The films were prepared by casting method and fully characterized by Scanning Electron Microscopy (SEM), Fourier-transform infrared spectroscopy (FTIR) and Differential Scanning Calorimetry (DSC). The grammage, thickness, contact angle, mechanical, optical and barrier properties of the films were measured, together with the evaluation of their biodegradability, antioxidant and antibacterial activities against common foodborne pathogens (*Enterococcus faecalis* and *Listeria monocytogenes*). The zein-based films with the incorporation of licorice essential oil presented the typical rugosities of the lotus leaf making the surfaces very hydrophobic (water contact angle of 112.50°). In addition to having antioxidant and antibacterial properties, the films also shown to be biodegradable, making them a strong alternative to the traditional plastics used in food packaging.

Keywords: lotus-effect; water contact angle; food packaging; licorice essential oil; antioxidant properties; antibacterial activity; foodborne pathogens

1. Introduction

Food packaging is designed to protect food from external factors, such as temperature, light or humidity that can lead to degradation [1]. Moreover, packages also protect its content from other environmental influences, namely, odors, microorganisms, shocks, dust, vibrations and compressive forces [1].

The production and application of synthetic materials in food packaging has grown quickly over the past few decades, resulting in serious environmental concerns due to the resistance to degradation of these synthetic materials [2,3]. In recent years, the replacement of synthetic plastics by natural polymers in packaging materials has been an intense research field [3,4]. Particularly, bio-based natural polymers (mainly polysaccharides and proteins) have been given increasing attention to be used in food packaging films because of their abundance, biodegradability, biocompatibility, and non-toxicity [3,4].

However, natural polysaccharides have their weak points in making hydrophobic films because most of them are hydrophilic and water absorbing, resulting in rapid solubilization in aqueous environments [5].

Protein-based films have several advantages over other types of edible materials. Proteins are constructed of nearly 20 different amino acids, and usually have good film forming capability [6]. Moreover, protein-based films are generally good gas barriers, possess good mechanical properties, and can also be regarded as nutrients [6].

Zein, a by-product obtained from corn starch processing, is prepared from corn protein flour. Zein is particularly rich in hydrophobic and neutral amino acids as well as some sulfur-containing amino acids, but lacks polar or ionizable amino acids [7]. Due to its large number of hydrophobic groups, zein is soluble in aqueous ethanol, yet insoluble in pure water. Since it possesses well-known film-forming ability caused by its unique amino acid composition, it is widely used in food packaging materials [7]. Pure zein films have good water barrier properties, but their mechanical properties are relatively poor. In order to overcome these deficiencies, blend zein films with other biodegradable biopolymers has been widely studied, and the zein provided good potential to produce blend films [7].

Superhydrophobic surfaces are used by several living organisms, both animals and plants. The most famous case from the plant kingdom (*Plantae*) is the lotus leaf (*Nelumbo nucifera*) [8]. The extreme water repellency of lotus leaves stems from the combination of low surface energy with the hierarchical topology present on the leaf surface [9]. Water droplets roll freely on these surfaces and remove dirt, keeping the leaves clean, even in the muddy waters where these plants tend to grow [9]. This extreme water repellency and self-cleaning performance of the lotus leaf is usually known as the "lotus effect" [10]. Recently, researchers have focused on recreating these surfaces, structured at the micro and nanometric scale, with new functionalities, replicating or mimicking the hierarchical surface morphology of the lotus leaf [11].

Licorice (*Glycyrrhiza glabra* L.) is a plant belonging to the *Fabaceae* family. Its sweet flavor makes it a popular ingredient in the production of candies and sweets in Europe [12]. The antioxidant and antibacterial properties of the licorice essential oil (EO) have already been demonstrated, together with its incorporation on carboxymethyl xylan films with potential to be used as novel food packaging materials [13].

Therefore, the aim of this work was to develop hydrophobic zein-based films incorporating licorice EO while biomimicking the lotus leaf surface, which is the innovation of this work. The films obtained were then characterized, and their bioactive properties evaluated.

2. Materials and Methods

2.1. Reagents

Polydimethylsiloxane (PDMS)—Sylgard® 184 Silicone Elastomer was obtained from Dow Corning (Midland, MI, USA). Zein from maize (CAS Number: 9010-66-6), presenting a molecular weight of 22–24 kDa, was supplied by Sigma-Aldrich (Saint Louis, MO, USA). Glycerol (anhydrous extra pure) was purchased from Merck (Darmstadt, Germany). The licorice EO (*Glycyrrhiza glabra* L.) (*Leguminosae*/*Fabaceae*) was obtained from Best Formula Industries (BF1, Kuala Lumpur, Malaysia). The EO (Pure Essential) was isolated from trunks of the plant by steam distillation and was stored at −20 °C in the dark until analysis and further use. The purity of the EO was tested and its quality ensured to be consistent with the standards described in the European Pharmacopoeia, being suitable to be used in products for human consumption. The chemical composition, analyzed by Gas Chromatography coupled to Mass Spectrometry (GC-MS, Perkin Elmer Clarus 600, Shelton, CT, USA), and the bioactivities, namely antioxidant ant antibacterial, of this EO were previously published [13].

2.2. Lotus Leaves

The fresh lotus leaves (*Nelumbo nucifera*) were collected in Autumn 2018 (September-October) and were kindly provided by the Jardim Botânico of Universidade de Coimbra. The hierarchical surface morphology of the lotus leaf was observed by Scanning Electron Microscopy (SEM, Hitachi, Chiyoda, Japan). For that, pieces of lotus leaves were fixed with 5% (*v*/*v*) glutaraldehyde at 4 °C for 12 h. After that, samples were washed once with phosphate buffer saline (PBS) and the dehydration was carried out in ethanol series for 1 h each (30, 50, 70, 80, 90% (*v*/*v*), and absolute) at room temperature [14,15]. The samples were then allowed to dry overnight in a desiccator. Finally, samples were coated with gold and analyzed by VP SEM Hitachi S-3400N (Hitachi, Chiyoda, Japan) using a voltage of 20.0 kV and 100.0 µA emission.

2.3. Negative Template Fabrication

PDMS was used to produce the lotus leaf negative template. The lotus leaves were gently cut and glued to glass Petri dishes. Then, molten paraffin was placed on the edges of the Petri dishes to seal the samples and prevent the polymer from dripping. Initially, the mixture of PDMS and its catalyzer (10:1; *w*/*w*) were weighed and mixed by hand for 5 min in order to improve homogeneity. Then, this mixture was degassed under vacuum until visible air bubbles formed during the mixing procedure had disappeared (30 min) [16]. Finally, the mixture was cast on the previously prepared Petri dishes with the lotus leaves [17]. The Petri dishes were then placed into a ventilated oven for a curing time of 4 h at 60 °C and then were left for more 48 h at room temperature [16] (Figure 1a). The negative template produced was visualized by SEM. Small pieces of the negative template were directly mounted on stubs and then coated with gold and analyzed by VP SEM Hitachi S–3400N using a voltage of 20.0 kV and 100.0 µA emission.

2.4. Preparation of Zein-Based Films

The film-forming solution was prepared by dissolving zein powder in 80% (*v*/*v*) ethanol-water solution (10%; *w*/*v*) under magnetic stirring for 30 min at 80 °C and 250 rpm. Glycerol at 25% (g glycerol/g dry zein powder) was added to the solution as plasticizer agent and stirred for 8 min at 80 °C and 250 rpm [18]. Then, licorice EO at 30% (g EO/g dry zein powder) was incorporated and was stirred again (8 min, 80 °C, 250 rpm) [13,18]. Films were then obtained by casting. For this purpose, the film-forming solution was spread on polystyrene Petri dishes in which the lotus leaf negative template was mounted as described above (Figure 1a), allowing the obtention of zein-based films with similar hierarchical surface morphology to that of the lotus leaf (positive replica). Films were also obtained by spreading the film-forming solution on simple polystyrene Petri dishes. Finally, the Petri dishes were placed into a ventilated oven to dry the mixture (24 h, 40 °C). The positive replica films obtained were observed by SEM as described for the negative template.

2.5. Characterization of Films

The control film (without EO) and the film with EO, both casted on simple polystyrene Petri dishes and on lotus negative template, were further characterized.

2.5.1. Fourier-Transform Infrared Spectroscopy (FTIR)

FTIR spectra of the films were obtained between 4000 and 600 cm^{-1} using a Nicolet iS10 smart iTRBasic (Thermo Fisher Scientific, Waltham, MA, USA) model, with 64 scans and a 4 cm^{-1} resolution [19].

2.5.2. Differential Scanning Calorimetry (DSC)

DSC analysis of the films was performed on a calorimeter (Netzsch DSC 204, Selb, Germany) operating in the following conditions: heating rate of 2 °C/min, inert atmosphere, and temperature range from 22 to 350 °C. For all the analyses, the respective baselines were firstly obtained [20].

2.5.3. Grammage, Thickness and Mechanical Properties

The grammage of the films was calculated based on the ratio between their mass and area (g/m^2), according to ISO 536:1995. The thickness (μm) was measured according to ISO 534:2011 using a micrometer (Adamel Lhomargy Model MI 20, Veenendaal, Netherlands), with several random measurements being considered. Tensile strength (MPa), elongation (%) and elastic modulus (MPa) of the films were obtained using a tensile tester (Thwing-Albert Instrument Co., West Berlin, NJ, USA), at 23 ± 2 °C and 50 ± 5% relative humidity (RH), as per ISO 1924/1 with a single modification. The initial grip was set at 50 mm and the crosshead was set at 10 mm/min [21].

2.5.4. Optical Properties

The color and transparency of the films were evaluated using a Technidyne Color Touch 2 spectrophotometer (New Albany, IN, USA). Measurements were performed on at least three random positions of the films using the D65 illuminant and 10° observer. Color coordinates L* (lightness), a* (redness; ±red-green) and b* (yellowness; ±yellow-blue) were obtained. The transparency of the films was calculated according to the equation defined in ISO 2289 [21].

2.5.5. Barrier Properties: Water Vapor Permeability

Water vapor permeability (WVP; g/(Pa.day.m)) and water vapor transmission rate (WVTR; $g/(m^2.day)$) were determined according to the standard protocol ASTM E96-00. The films, which had been equilibrated at 23 ± 2 °C and 50 ± 5% RH for 72 h, were fixed on the top of equilibrated cups containing a desiccant (15 g of anhydrous $CaCl_2$, dried at 105 °C for 2 h before being used). The test cups were placed in a cabinet at 23 ± 2 °C and 50 ± 5% RH. The weight changes were monitored for every 2 h over 48 h. The gradient was calculated from the slope of a linear regression of the weight increase versus time [22,23]. WVTR and WVP were calculated according to the following equations:

$WVTR = (\Delta m/\Delta t)/A$, where Δm is the weight changes of test cups (g), A is the test area (m^2) and t is test time (day).

$WVP = WVTR/\Delta p = (WVTR/[p(RH_1 - RH_2)]) \times e$, where p is the vapor pressure of water at 23 °C (Pa), RH_1 is the RH of the cabinet (50%), RH_2 is the RH inside the cups (0%) and e is the thickness (m) of the films.

2.5.6. Contact Angle Measurement

Contact angle measurements were performed using the sessile drop method with distilled water as test liquid (OCAH 200, DataPhysics, Filderstadt, Germany). At least eight measurements were made for each film [21]. Moreover, the contact angle was measured for both sides (upper and lower) of each film.

2.5.7. Water Solubility

The films solubility in water was defined by the content of dry matter solubilized after 24 h of immersion in water. The initial dry matter content of each film was determined by drying to constant weight in an oven at 105 °C. Two disks of films (2 cm of diameter) were cut, weighed and immersed in 50 mL of water. After 24 h of immersion at 20 °C with occasional magnetic stirring, the pieces of films were taken out (by filtration) and dried to constant weight in an oven at 105 °C, in order to determine the weight of dry matter which was not solubilized in water [24]. The measurement of solubility of the films was determined as follows:

$S(\%) = [(mi - mf)/(mi)] \times 100\%$, where S is the solubility, mi is the initial mass and mf the final mass.

2.5.8. Antioxidant Activity

The antioxidant activity of the films was evaluated by the 2,2-diphenyl-1-picrylhydrazyl (DPPH) free radical scavenging assay and the β-carotene bleaching test.

For the DPPH free radical scavenging assay, 3 disks of the film (6 mm of diameter) were added to 2.9 mL of a DPPH methanolic solution (0.1 mM). Then, the absorbances were measured at 517 nm every 30 min, for 5 h, against a blank of methanol [13].

For the β-carotene bleaching test, 500 µL of a β-carotene solution (20 mg/mL in chloroform) were added to 40 µL of linoleic acid, 400 µL of Tween 40 and 1 mL of chloroform. The chloroform was then evaporated under vacuum, and 100 mL of oxygenated distilled water were added to the mixture to form an emulsion. Then, 5 mL of this emulsion were pipetted into test tubes containing 3 disks of the film (6 mm of diameter). Finally, the tubes were shaken and placed at 50 °C in a water bath for 1 h. The absorbances of the samples were measured at 470 nm [13].

2.5.9. Antibacterial Properties

The antibacterial properties of the films against two foodborne pathogens (*Enterococcus faecalis* ATCC 29212 and *Listeria monocytogenes* LMG 16779) were evaluated by solid diffusion assay. For this test, inoculums were prepared by suspending bacteria in a sterile saline solution (NaCl; 0.85%; w/v) to a cell suspension of 0.5 McFarland (1–2 × 10^8 colony-forming units/mL (CFU/mL)). Disks of the films (6 mm of diameter) were prepared under aseptic conditions. Then, the Müeller-Hinton agar (MHA) plates were inoculated and allowed to dry, being the disks of films placed over the inoculated culture medium. Finally, the plates were incubated at 37 °C for 24 h. After the incubation, all the plates were visually checked for inhibition zones, being their diameters measured using a pachymeter. This assay was performed three independent times [13,21].

2.5.10. Biodegradability

Soil burial degradation test was carried out to evaluate the biodegradability of the films. Small pieces of films (dimensions 2 × 7 cm) were buried in natural soil at a depth of 10 cm (23 ± 2 °C and 50 ± 5% RH). After 10 days, the samples were collected, washed several times with distilled water and then allowed to dry in an oven at 50 °C for 24 h [25]. The weight loss was monitored and the FTIR spectra of the degraded films were also recorded as described above.

2.6. Statistical Analysis

The results were expressed as mean ± standard deviation (SD). The data were analyzed using the statistical program IBM SPSS Statistics 25 (https://www.ibm.com/analytics/spss-statistics-software). The significant difference among means was analyzed by Student's T-test (assuming the normal distribution of the continuous variables). A level of p-value < 0.05 was considered significant. Linear regression was also performed for antioxidant activity of the films measured by DPPH scavenging assay considering the % Inhibition as a function of time.

3. Results and Discussion

3.1. Proof-of-Concept: Negative Template and Positive Replica

This work intended to develop zein-based films incorporating licorice EO with potential applications in food packaging materials. Considering that these bio-based films are usually hydrophilic, and considering the "lotus effect", it was decided to try to mimic the hierarchical surface morphology of the lotus leaves in these films.

Firstly, the surface of the leaves was analyzed by SEM (Figure 1b), where it was possible to notice the common pattern of lotus leaves with its micro and nanostructures, measuring between 7 and 8 µm. Then, to prepare a negative template of the lotus leaves, a mixture of PDMS and its catalyzer was nano-casted on fresh lotus leaves. After solidification of PDMS, the negative template was lift off and visualized by SEM (Figure 1c). The PDMS-negative template presents the holes corresponding to the structures observed in the lotus leaves surface, which means that a complementary topographic surface structure of the original template (lotus leaves) was successfully obtained. Finally, the zein-based films (with and without licorice EO) were casted on the PDMS-negative template to replicate the hierarchical surface morphology of the lotus leaves in the films. The films obtained were once again visualized by SEM (Figure 1d), ant it was possible to observe that the complex surface patterns of the lotus leaves were transferred to the surface of the zein-based films with high fidelity (positive replica), which validated the original concept behind this study. Moreover, the structures obtained in both types of zein-based films presented similar dimensions (≈7–8 μm), measured by SEM, indicating that the incorporation of the licorice EO in the films did not interfere with the nano-casting process. Other researchers had previously replicated the so-called lotus-leaf-like structures for biomedical and pharmaceutical applications [10,26–28].

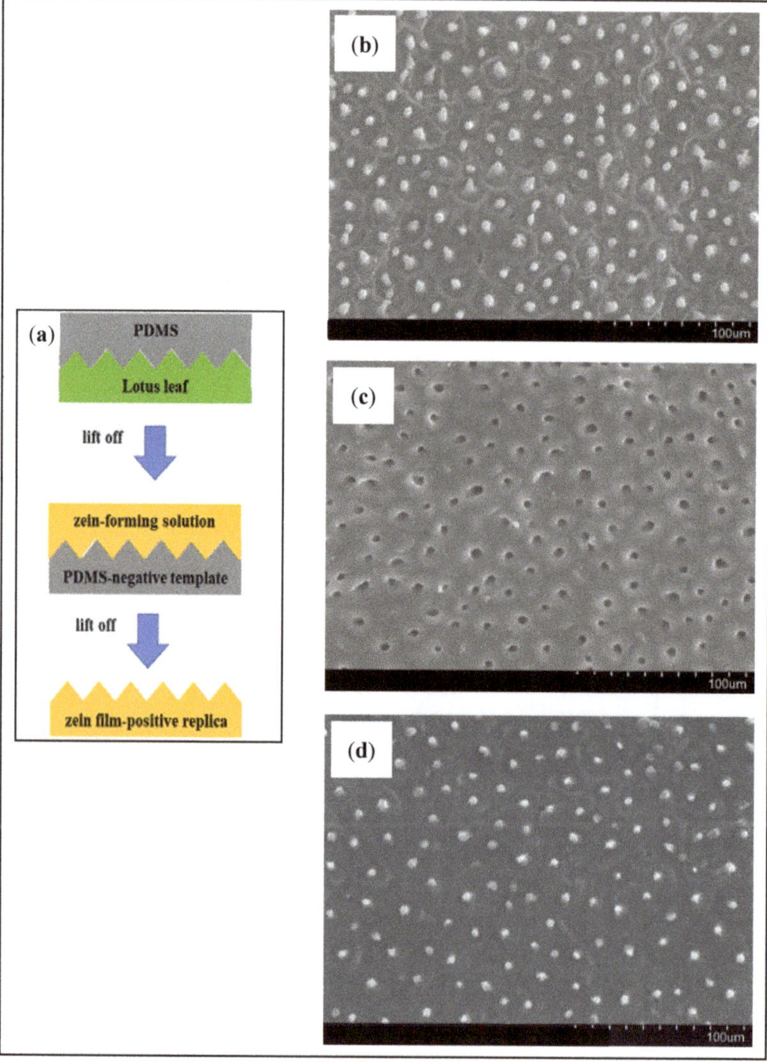

Figure 1. Flowchart for negative template fabrication and films preparation (**a**). SEM images of lotus leaf surface (**b**), PDMS-negative template (**c**), and zein film-positive replica (**d**). (Magnification: 500×).

3.2. FTIR and DSC Analyses of The Films

The spectra of zein-based films with or without licorice EO, and both casted on simple polystyrene Petri dishes and on lotus negative template were recorded. It was observed that all the films without incorporation of licorice EO presented analogous FTIR spectra. Similar results were found for the films with licorice EO, indicating that the support material where the films were casted (polystyrene or PDMS) did not influence the chemical interactions between the components of the films. In addition, both sides of each film presented similar FTIR spectra, which means that the topographic surface structure did not affect the chemical composition of the films. Figure 2a shows the FTIR spectrum of the zein film without licorice EO, while Figure 2b presents the spectrum of the film incorporating the EO. The licorice EO is mainly constituted (71.8%) by isopropyl palmitate [13], also shown in Figure 2b. The large absorption at 2960 cm^{-1} in the zein film with licorice EO was attributed to the stretching vibration of the aliphatic compounds, and the peak at 1740 cm^{-1} was ascribed to the stretching vibration of the ester compounds. Isopropyl palmitate possesses an aliphatic chain with several methyl groups (CH), while being also an ester presenting a carbonyl group (C = O) [13]. These molecular features of the main compound of the EO are responsible for the peaks visible in the FTIR spectrum of the zein-based film incorporating licorice EO, when compared to the spectrum of the control film, indicating that the EO was successfully incorporated.

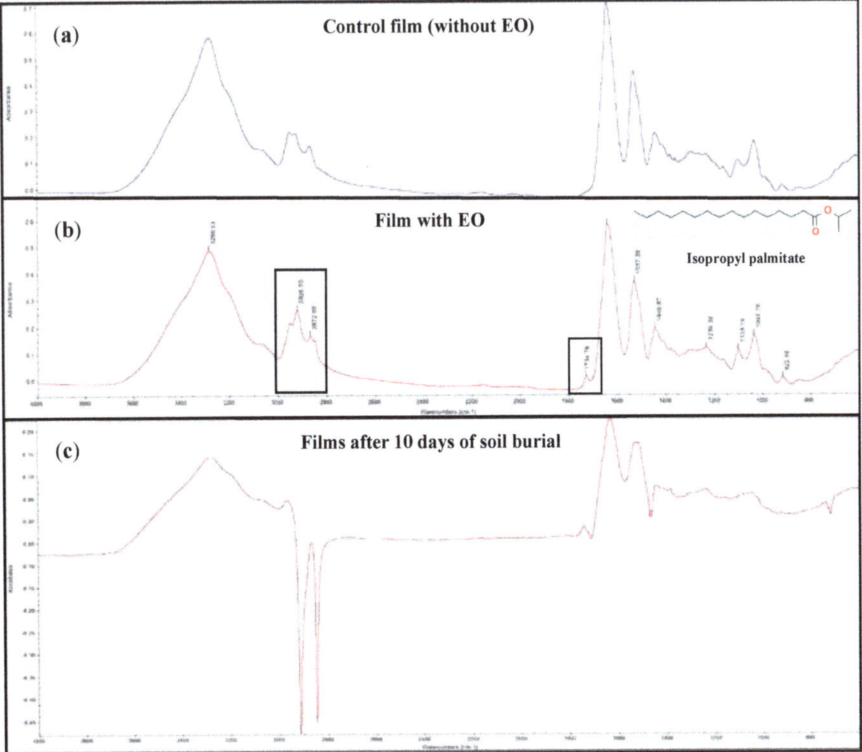

Figure 2. FTIR spectra of the films: control film (without EO) (**a**), film with EO (**b**), and films after 10 days of soil burial (**c**).

The thermal profiles of zein films with or without the licorice EO incorporated were evaluated by DSC. Similar to what was observed in FTIR analysis, the support material where the films were casted did not influence the DSC curves. Figure 3a shows the DSC curve of a zein film without EO and Figure 3b shows the DSC curve of a zein-based film incorporating licorice EO. Both thermograms show endothermic peaks in the range of temperature of 50 °C to 150 °C. The presence of such peaks is attributed to the loss of volatile components, like water, or the possibility of chain relaxation. Additionally, in this range of temperature, the breakdown of hydrogen bonds that are present in the zein structure and other molecular associations also happens [20]. Proteins have some features associated to their different tridimensional structure, such as the denaturation process. The DSC curve of zein-based film without EO (Figure 3a) exhibited endothermic peaks at 311 °C and 318 °C, which can be interpreted as the protein unfolding, similar to what was previously observed [20]. Moreover, through the DSC curve of the control film, at the 174 °C mark, a slight modification on the linear profile of the zein curve appears, which is linked to the zein glass transition temperature (Tg), since above this temperature the protein chains of zein enter in a flexible stage [20]. When incorporating licorice EO in the zein films, the DSC curve (Figure 3b) changed, having observed a lower value of Tg (165 °C), which is probably related to the loss of some flexibility in the films with the EO [20]. This fact can be explained by the interactions among zein and the compounds of licorice EO after the preparation of the films, as it was also observed by the FTIR analysis. Similar results were previously obtained by other authors dealing with zein-based films [29].

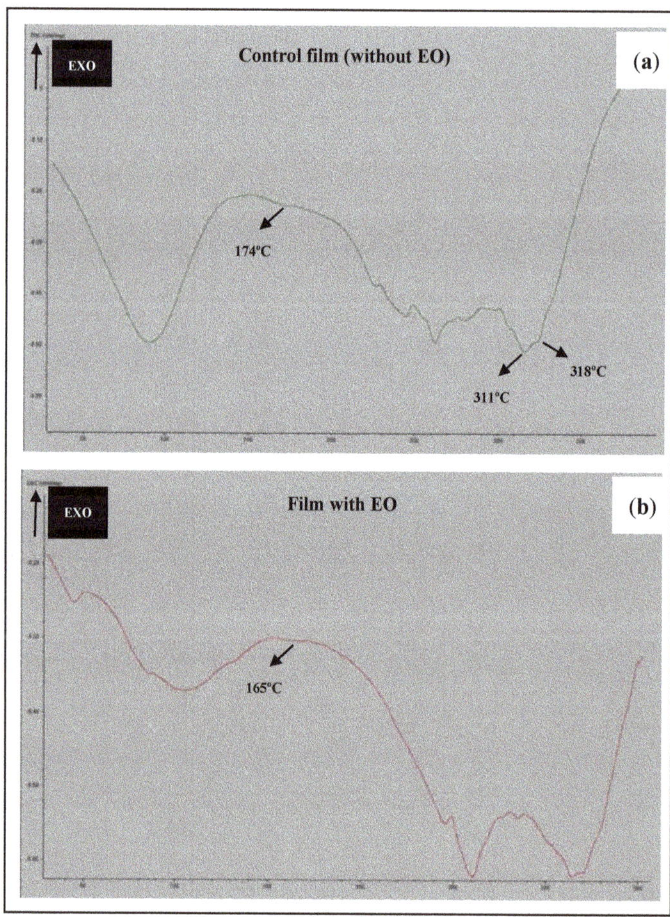

Figure 3. DSC curves of the films: control film (without EO) (**a**), and film with EO (**b**).

3.3. *Grammage, Thickness, Mechanical and Optical Properties of The Films*

The grammage and thickness were significantly higher (p-value < 0.05) in the films incorporating licorice EO casted on the lotus negative template (Table 1), which is probably due to the imprisonment of air microbubbles during the casting. The film-forming solution must fill all the holes of the lotus negative template, which are full of air that can become entrapped in the film matrix. Moreover, the incorporation of the EO in the films most likely increases their apparent density, which is also reflected in higher values of grammage.

Table 1. Grammage, thickness, mechanical and optical properties of the films.

Properties	Films				p-Values
	Polystyrene Petri Dishes		Lotus Negative Template		
	Without EO [a]	With EO [b]	Without EO [c]	With EO [d]	
Grammage (g/m^2)	107.30 ± 5.33	145.10 ± 18.64	128.55 ± 3.35	167.21 ± 1.41	0.088 [ab] **0.023** [cd] **0.020** [ac] 0.176 [bd]
Thickness (µm)	118.70 ± 10.70	119.33 ± 13.92	156.67 ± 29.66	166.67 ± 10.99	0.914 [ab] 0.460 [cd] **0.024** [ac] **< 0.001** [bd]
Tensile strength (MPa)	11.20 ± 2.08	9.60 ± 0.55	6.82 ± 0.57	5.31 ± 0.62	0.054 [ab] 0.063 [cd] **0.022** [ac] **0.001** [bd]
Elongation (%)	2.51 ± 0.36	2.44 ± 0.36	3.13 ± 0.22	2.21 ± 0.38	0.823 [ab] **0.032** [cd] 0.076 [ac] 0.489 [bd]
Elastic modulus (MPa)	927.12 ± 110.02	353.21 ± 20.21	172.49 ± 25.49	257.97 ± 31.80	0.111 [ab] 0.177 [cd] 0.079 [ac] 0.149 [bd]
Transparency (%)	94.89 ± 0.45	63.78 ± 2.56	82.38 ± 0.98	69.92 ± 1.56	0.050 [ab] **0.033** [cd] **0.037** [ac] 0.203 [bd]
L* (lightness)	25.99 ± 0.47	62.35 ± 1.65	46.67 ± 1.41	56.85 ± 0.44	**0.019** [ab] 0.066 [cd] **0.027** [ac] 0.167 [bd]
a* (redness)	−1.43 ± 0.07	0.04 ± 0.39	0.39 ± 0.05	2.69 ± 1.01	0.158 [ab] 0.263 [cd] **0.003** [ac] 0.199 [bd]
b* (yellowness)	10.74 ± 1.95	37.70 ± 1.51	26.49 ± 1.70	42.73 ± 2.31	**0.010** [ab] **0.036** [cd] **0.027** [ac] 0.230 [bd]

(Results expressed as mean ± SD) a, b, c and d correspond to each type of film; Significant p-values are highlighted in bold.

Concerning the mechanical properties, zein films are brittle, and thus, plasticizers are needed to improve their flexibility [30]. In general, tensile strength and elongation are key factors in packaging materials, which keep their integrity during packaging, transport, storage and sales [7]. The tensile strength of the films casted on lotus negative template was significantly lower ($p < 0.05$) when compared to the films casted in simple polystyrene Petri dishes (Table 1), incorporating (or not) licorice EO. This means that it is not the licorice EO that affects tensile strength, but the support material where the films were casted, probably because of the same effect of air microbubble entrapment, as explained above.

Nevertheless, elongation and elastic modulus were similar within all the films (Table 1). Finally, the mechanical properties of the zein-based films developed in this work, using glycerol as plasticizer, are very close to the ones obtained previously for zein films using oleic acid as the plasticizer agent [29,30], but improved upon than the ones obtained using other plasticizers (buriti oil, macadamia oil and olive oil) [29].

The optical properties (L*, a* and b*) of the films with EO were only slightly affected (Table 1), since zein is a yellow powder and the licorice EO is yellow-brownish. On a previous work, the licorice EO was incorporated in carboxymethyl xylan films (white) and the optical properties were significantly changed by the EO, particularly transparency [13]. In the present work, the color of the EO is not a problem since zein is also yellow.

3.4. Barrier Properties: Water Vapor Permeability

The barrier properties of the films, and particularly water permeability, plays a critical role in evaluating the practical applications of functional films. It was expected that the films with licorice EO presented lower water permeability since the hydrophobic nature of the EO was known to contribute to the decrease of WVP [22]. Furthermore, in the present work, the WVP was significantly higher (p-value < 0.05) in the zein films casted on lotus negative template, compared to the ones casted on simple polystyrene Petri dishes (Table 2). The WVP of the films casted on polystyrene is near 2.74×10^{-6} g/(Pa.day.m) while for the films casted on lotus negative template it is about $3–4 \times 10^{-6}$ g/(Pa.day.m), indicating that the water barrier properties of the films were weakened. The support material where the films were casted seems to influence the WVP, probably because of the higher values of thickness obtained for the films casted on lotus negative template. The zein-based films developed in the present work presented strong water barrier properties than the ones obtained previously with chitosan-zein edible films incorporated with anise, orange and cinnamon essential oils (250 ppm) [31].

Table 2. Barrier properties of the films: water vapor permeability.

Properties	Films				p-Values
	Polystyrene Petri Dishes		Lotus Negative Template		
	Without EO [a]	With EO [b]	Without EO [c]	With EO [d]	
WVTR (g/(m^2.day))	31.53 ± 3.38	30.33 ± 3.04	27.23 ± 0.68	27.23 ± 4.73	0.672 [ab] 0.991 [cd] 0.156 [ac] 0.402 [bd]
WVP (g/(Pa.day.m)) ($\times 10^{-6}$)	2.74 ± 0.167	2.74 ± 0.274	3.58 ± 0.049	4.46 ± 0.178	1.000 [ab] **0.009** [cd] **0.009** [ac] **0.002** [bd]

(Results expressed as mean ± SD) a, b, c and d correspond to each type of film; Significant p-values are highlighted in bold.

3.5. Contact Angle and Water Solubility

Understanding the wettability of the films is often carried out by measuring the contact angle formed between a liquid drop and the film [32]. Additionally, the water contact angle measurements provide information about the hydrophobicity/hydrophilicity of the surface of the films. A hydrophobic surface is a surface in which the water contact angle is higher than 90° [5]. Table 3 summarizes the water contact angle measured in both sides of the films (with and without the licorice EO), casted on polystyrene and on lotus negative template. The addition of licorice essential oil increases significantly

(p-values < 0.05) the contact angle of the films (Table 3). In films casted on simple polystyrene Petri dishes, the contact angle raises from 48.27° (without EO) to 71.80° (with EO). These results may be explained by the chemical composition of the licorice EO, mainly constituted by isopropyl palmitate, which presents an octanol/water partition coefficient (logP) of 8.16, an indicator of its strong hydrophobic character [13]. Licorice is usually taken as a hydrophobic component of edible films and coatings [33]. More interestingly was the effect observed in the films casted on lotus negative template. The upper side of the films (without the hierarchical surface morphology of the lotus leaves), presented contact angles of 65.15° (without EO) and 58.04° (with EO); contrariwise to what was verified in the lower side of the films (with the hierarchical surface morphology of the lotus leaves) that presented significantly higher (p-value < 0.05) contact angles than the ones obtained for the films casted on polystyrene (Table 3). The film without EO that presents the complex surface patterns of the lotus leaves (positive replica) had a contact angle of 81.95°; significantly higher (p-value < 0.05) contact angle was obtained for the film with EO, which is also a lotus positive replica (112.50°) (Table 3). Here, the strong hydrophobic effect of licorice/isopropyl palmitate was noticed, in addition to the "lotus effect", which confirms the initial idea behind this work. Combining the "lotus effect" with the incorporation of the licorice EO resulted in a near superhydrophobic surface (water contact angle superior than 150°) [5]. To the best of our knowledge, it was the first time that zein-based films were obtained with such high water contact angle.

Water solubility is a measure of the resistance of the films to water [24]. The results obtained showed that all the films presented a water solubility of about 20% (Table 3), indicating the lower affinity to water of these films, which also corroborates the water contact angles obtained for the films.

3.6. Antioxidant and Antibacterial Properties

Figure 4 shows the antioxidant activity of the films measured by DPPH scavenging assay over time. All the films presented the ability to scavenge the free radicals in a time-dependent manner, as demonstrated by the linear regressions (p-values < 0.05). After 5 h of reaction, the films presented near 90% of inhibition of DPPH free radicals (Figure 4). Although the licorice EO is a potent antioxidant [13], the films without it also presented the capacity to scavenge free radicals, which is due to zein. The antioxidative nature of zein was previously reported, being frequently used in food packaging (edible films and coatings) without adding any other antioxidant during processing [34]. Some studies have reported the benefits of using zein films as packaging material for cooked turkey and fresh broccoli [34]. The capacity of the films to inhibit lipid peroxidation was also evaluated (Table 4). The results showed that the films with licorice EO have significantly higher (p-value < 0.05) percentages of inhibition than the control films. Both types films (polystyrene and lotus negative template) incorporating the EO presented similar capacity to inhibit the lipid peroxidation measured by the β-carotene bleaching test (20–30%), contrariwise to what was obtained with the control films (3–5%) (Table 4). This capacity can be attributed to isopropyl palmitate, the major compound of licorice EO, which is generally used in cosmetic and food industries due to its binder and fragrant properties, together with skin-conditioning and emollient activities. The Food and Drug Administration (FDA) considers that this compound is not ecotoxic and classifies it as not expected to be potentially toxic or harmful, presenting a low human health priority [35].

Also, licorice is generally recognized as safe (GRAS) by the FDA, indicating that there is no evidence, in the available information on licorice, that identifies a hazard to the public when it is used at levels that are now current and in the manner now practiced [36].

Table 3. Contact angle, water solubility and weight loss of the films.

Properties	Films								*p*-Values
	Polystyrene Petri Dishes				Lotus Negative Template				
	Without EO [A]		With EO [B]		Without EO [C]		With EO [D]		
	Lower Side [a]	Upper Side	Lower Side [b]	Upper Side	Lower Side (Lotus Replica) [c]	Upper Side [d]	Lower Side (Lotus Replica) [e]	Upper Side [f]	
Contact angle (°)	48.27 ± 3.76	31.49 ± 1.05	71.80 ± 8.60	69.92 ± 3.02	81.95 ± 8.74	65.15 ± 3.11	112.50 ± 3.48	58.04 ± 5.71	**0.009** [ab] **< 0.001** [ce] **< 0.001** [ac] **0.001** [be] **0.001** [cd] **< 0.001** [ef]
Water solubility (%)	20.33 ± 1.45		23.41 ± 2.18		22.38 ± 0.73		22.38 ± 1.57		0.118 [AB] 0.546 [CD] 0.122 [AC] 1.000 [BD]
Weight loss (%)	61.97 ± 3.16		56.61 ± 1.18		63.72 ± 2.82		47.66 ± 3.31		0.181 [AB] **0.003** [CD] 0.955 [AC] **0.031** [BD]

(Results expressed as mean ± SD) a, b, c, d, e and f correspond to each type of film (and side); A, B, C and D correspond to the films with or without the EO; Significant *p*-values are highlighted in bold.

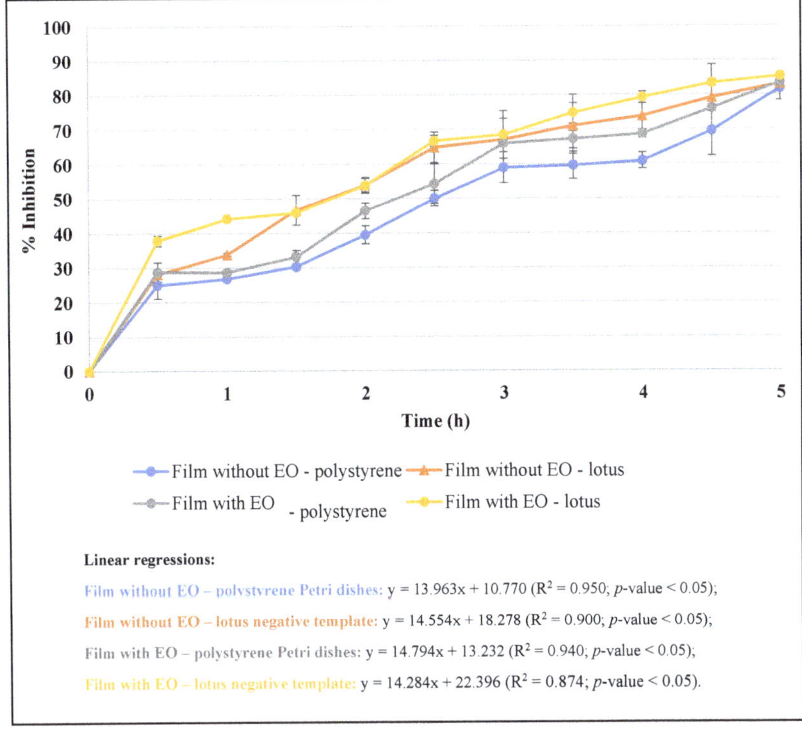

Figure 4. Antioxidant activity of the films measured by DPPH scavenging assay.

Table 4. Antioxidant activity (β-carotene bleaching assay) and diameters of inhibition zones obtained with the films.

Properties	Films				*p*-values
	Polystyrene Petri Dishes		Lotus Negative Template		
	Without EO [a]	With EO [b]	Without EO [c]	With EO [d]	
% Inhibition (β-carotene bleaching test)	3.45 ± 0.26	29.00 ± 1.67	5.91 ± 0.64	22.36 ± 1.29	**<0.001** [ab] **<0.001** [cd] **0.011** [ac] **0.007** [bd]
Diameters of inhibition zones (mm)	**Without EO** [a]	**With EO** [b]	**Without EO** [c]	**With EO** [d]	***p*-values**
E. faecalis ATCC 29219	6.45 ± 0.24	7.68 ± 0.39	6.52 ± 0.36	7.74 ± 0.40	**0.015** [ab] **0.018** [cd] 0.795 [ac] 0.861 [bd]
L. monocytogenes LMG 16779	6.58 ± 0.33	8.25 ± 0.42	6.64 ± 0.35	8.46 ± 0.39	**0.007** [ab] **0.004** [cd] 0.840 [ac] 0.560 [bd]

(Results expressed as mean ± SD) a, b, c and d correspond to each type of film; Significant *p*-values are highlighted in bold.

The antibacterial properties of the films were evaluated against two well-known foodborne pathogens (*E. faecalis* ATCC 29212 and *L. monocytogenes* LMG 16779). *E. faecalis* is a commensal of the human gastrointestinal tract that can persist in the external environment and is a leading cause of several infections. Given its diverse habitats, the organism has developed numerous strategies to survive a multitude of environmental conditions [37]. *L. monocytogenes* is a foodborne pathogen responsible for a disease called listeriosis, which is potentially lethal in immunocompromised people and can provoke septicemia, meningitis and fetal infection or abortion in infected pregnant women [38].

The results of the antibacterial activity studied by solid diffusion assay are presented in Table 4. Both types of zein films incorporating licorice EO presented significantly higher (p-value < 0.05) diameters of inhibition zones for the two bacterial species. Moreover, the anti-biofilm potential of zein films against the same foodborne pathogens was evaluated by SEM, forming the bacterial biofilms directly on the surface of the films (results not shown). It was possible to verify that in the films presenting the hierarchical surface morphology of the lotus leaves, the bacterial adhesion did not occur. More than inhibiting the bacterial growth, the films inhibited the adhesion. The anti-bio adhesion of surfaces with lotus-leaf-like rugosities is well described in the literature [39].

Since zein-based films incorporating licorice EO were able to scavenge free radicals to inhibit lipid peroxidation and the growth of foodborne pathogens, they can potentially be used as alternative food packaging systems, particularly in foods with high contents of lipids.

3.7. Biodegradability

The biodegradability of the films was studied by soil burial degradation test for 10 days. The weight loss of zein-based films was about 50–60% (Table 3), which is a clear reflection of the biodegradation process performed by the microorganisms and moisture present in the soil. Moreover, at the end of the biodegradability test, the films were thin and appeared disrupt (Figure 5b) when compared to the initial samples (Figure 5a). The FTIR spectra of the films after the soil burial degradation test (Figure 2c) showed peaks absorption and decrease in intensities as biodegradation took place, as other authors also reported [25]. These results clearly show the biodegradable nature of the zein-based films now developed.

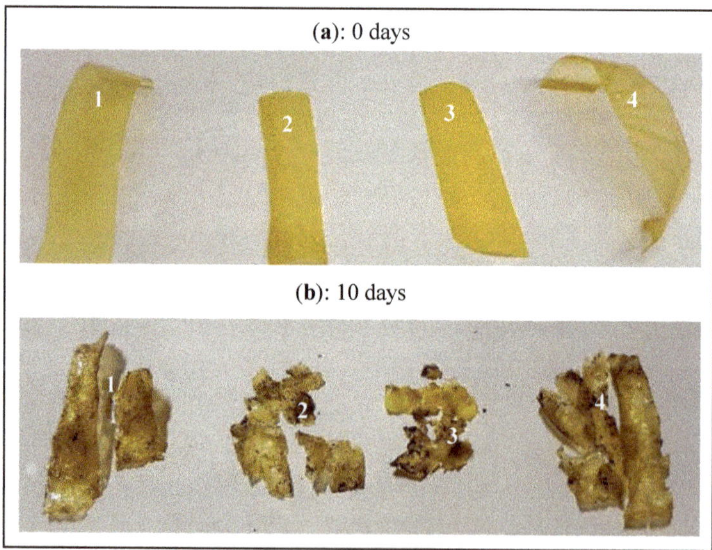

Figure 5. Biodegradability: soil burial degradation test of the films. Film without EO — polystyrene Petri dishes (1), film without EO — lotus negative template (2), film with EO —polystyrene Petri dishes (3), and film with EO —lotus negative template (4); initial films at 0 days (**a**), and films after soil burial degradation test (10 days) (**b**).

4. Conclusions

In this work, a simple and rapid method to mimic the lotus leaf surface was developed and applied to the production of zein-based films. The licorice EO was also incorporated into the films as a bioactive agent. The zein films produced using the lotus negative template presented lotus-leaf-like rugosities, resulting in very hydrophobic surfaces (water contact angle of 112.50°). The zein films with licorice essential oil are biodegradable and possess antioxidant and antibacterial properties against known foodborne pathogens, making them potential alternatives to the conventional plastics used in food packaging solutions, reducing environmental pollution and increasing the shelf-life of foods.

Future research is needed to identify ways to produce on a large scale films with the hierarchical surface morphology of the lotus leaf.

Author Contributions: Conceptualization, A.R.; Formal analysis, F.D. and A.R.; Investigation, Â.L. and A.R.; Methodology, Â.L. and A.R.; Resources, F.D. and A.R.; Supervision, F.D.; Validation, Â.L., F.D. and A.R.; Writing—original draft, Â.L.; Writing—review & editing, F.D. and A.R.

Funding: Ângelo Luís acknowledges the contract of Scientific Employment in the scientific area of Microbiology financed by Fundação para a Ciência e a Tecnologia (FCT). This work was partially supported by CICS-UBI that is financed by National Funds from FCT (UID/Multi/00709/2019).

Acknowledgments: Authors acknowledge Carine Azevedo (Jardim Botânico da Universidade de Coimbra) for kindly providing the lotus leaves. The authors would also like to thank to Ana Paula Gomes from the Laboratório de Microscopia Eletrónica (Universidade da Beira Interior) for her help in SEM and DSC assays.

Conflicts of Interest: The authors declare no conflict of interest.

References

1. Ribeiro-Santos, R.; Andrade, M.; Melo, N.R.; Sanches-Silva, A. Use of essential oils in active food packaging: Recent advances and future trends. *Trends Food Sci. Technol.* **2017**, *61*, 132–140. [CrossRef]
2. Cazón, P.; Velazquez, G.; Ramírez, J.A.; Vázquez, M. Polysaccharide-based films and coatings for food packaging: A review. *Food Hydrocoll.* **2017**, *68*, 136–148. [CrossRef]
3. Cao, T.L.; Song, K.B. Active gum karaya/Cloisite Na^+ nanocomposite films containing cinnamaldehyde. *Food Hydrocoll.* **2019**, *89*, 453–460. [CrossRef]
4. Yang, Y.C.; Mei, X.W.; Hu, Y.J.; Su, L.Y.; Bian, J.; Li, M.F.; Peng, F.; Sun, R.C. Fabrication of antimicrobial composite films based on xylan from pulping process for food packaging. *Int. J. Biol. Macromol.* **2019**, *134*, 122–130. [CrossRef] [PubMed]
5. Dong, F.; Padua, G.W.; Wang, Y. Controlled formation of hydrophobic surfaces by self-assembly of an amphiphilic natural protein from aqueous solutions. *Soft Matter* **2013**, *9*, 5933–5941. [CrossRef]
6. Tsai, M.J.; Weng, Y.M. Novel edible composite films fabricated with whey protein isolate and zein: Preparation and physicochemical property evaluation. *LWT-Food Sci. Technol.* **2019**, *101*, 567–574. [CrossRef]
7. Sun, H.; Shao, X.; Jiang, R.; Shen, Z.; Ma, Z. Mechanical and barrier properties of corn distarch phosphate-zein bilayer films by thermocompression. *Int. J. Biol. Macromol.* **2018**, *118*, 2076–2081. [CrossRef] [PubMed]
8. Hönes, R.; Rühe, J. Extending the Lotus Effect: Repairing Superhydrophobic Surfaces after Contamination or Damage by CHic Chemistry. *Langmuir* **2018**, *34*, 8661–8669. [CrossRef]
9. Choi, Y.; Brugarolas, T.; Kang, S.M.; Park, B.J.; Kim, B.S.; Lee, C.S.; Lee, D. Beauty of lotus is more than skin deep: Highly buoyant superhydrophobic films. *ACS Appl. Mater. Interfaces* **2014**, *6*, 7009–7013. [CrossRef]
10. Latthe, S.S.; Terashima, C.; Nakata, K.; Fujishima, A. Superhydrophobic surfaces developed by mimicking hierarchical surface morphology of lotus leaf. *Molecules* **2014**, *19*, 4256–4283. [CrossRef]
11. Saison, T.; Peroz, C.; Chauveau, V.; Berthier, S.; Sondergard, E.; Arribart, H. Replication of butterfly wing and natural lotus leaf structures by nanoimprint on silica sol-gel films. *Bioinspiration Biomim.* **2008**. [CrossRef]
12. Luís, Â.; Domingues, F.; Pereira, L. Metabolic changes after licorice consumption: A systematic review with meta-analysis and trial sequential analysis of clinical trials. *Phytomedicine* **2018**, *39*, 17–24. [CrossRef]
13. Luís, Â.; Pereira, L.; Domingues, F.; Ramos, A. Development of a carboxymethyl xylan film containing licorice essential oil with antioxidant properties to inhibit the growth of foodborne pathogens. *LWT-Food Sci. Technol.* **2019**, *111*, 218–225. [CrossRef]

14. Ensikat, H.J.; Ditsche-Kuru, P.; Barthlott, W. Scanning electron microscopy of plant surfaces: Simple but sophisticated methods for preparation and examination. *Microsc. Sci. Technol. Appl. Educ.* **2010**, 248–255.
15. Santos, C.; Teixeira, G.; Monteiro, A. Leaf morphoanatomy of Portuguese autoctones white grapevine cultivars of different geographical origin. *Xth Int. Terroir Congr.* **2014**, 103–108.
16. Hopf, R.; Bernardi, L.; Menze, J.; Zündel, M.; Mazza, E.; Ehret, A.E. Experimental and theoretical analyses of the age-dependent large-strain behavior of Sylgard 184 (10:1) silicone elastomer. *J. Mech. Behav. Biomed. Mater.* **2016**, *60*, 425–437. [CrossRef]
17. Xi, J.; Jiang, L. Biomimic superhydrophobic surface with high adhesive forces. *Ind. Eng. Chem. Res.* **2008**, *47*, 6354–6357. [CrossRef]
18. Kashiri, M.; Cerisuelo, J.; Domínguez, I.; López-Carballo, G.; Muriel-Gallet, V.; Gavara, R.; Hernández-Muñoz, P. Zein films and coatings as carriers and release systems of Zataria multiflora Boiss. essential oil for antimicrobial food packaging. *Food Hydrocoll.* **2017**, *70*, 260–268.
19. Nunes, M.R.; Castilho, M.S.M.; Veeck, A.P.L.; Rosa, C.G.; Noronha, C.M.; Maciel, M.V.O.B.; Barreto, P.M. Antioxidant and antimicrobial methylcellulose films containing Lippia alba extract and silver nanoparticles. *Carbohydr. Polym.* **2018**, *192*, 37–43. [CrossRef]
20. Müller, V.; Piai, J.F.; Fajardo, A.R.; Fávaro, S.L.; Rubira, A.F.; Muniz, E.C. Preparation and characterization of zein and zein-chitosan microspheres with great prospective of application in controlled drug release. *J. Nanomater.* **2011**. [CrossRef]
21. Silva, Â.; Duarte, A.; Sousa, S.; Ramos, A.; Domingues, F.C. Characterization and antimicrobial activity of cellulose derivatives films incorporated with a resveratrol inclusion complex. *LWT-Food Sci. Technol.* **2016**, *73*, 481–489. [CrossRef]
22. Chu, Y.; Xu, T.; Gao, C.C.; Liu, X.; Zhang, N.; Feng, X.; Liu, X.; Shen, X.; Tanh, X. Evaluations of physicochemical and biological properties of pullulan-based films incorporated with cinnamon essential oil and Tween 80. *Int. J. Biol. Macromol.* **2019**, *122*, 388–394. [CrossRef]
23. Queirós, L.C.C.; Sousa, S.C.L.; Duarte, A.F.S.; Domingues, F.C.; Ramos, A.M.M. Development of carboxymethyl xylan films with functional properties. *J. Food Sci. Technol.* **2017**, *54*, 9–17. [CrossRef]
24. Casariego, A.; Souza, B.W.S.; Cerqueira, M.A.; Teixeira, J.A.; Cruz, J.A.; Díaz, R.; Vicente, A.A. Chitosan/clay films' properties as affected by biopolymer and clay micro/nanoparticles' concentrations. *Food Hydrocoll.* **2009**, *23*, 1895–1902. [CrossRef]
25. Kumar, D.; Kumar, P.; Pandey, J. Binary grafted chitosan film: Synthesis, characterization, antibacterial activity and prospects for food packaging. *Int. J. Biol. Macromol.* **2018**, *115*, 341–348. [CrossRef]
26. Kim, S.I.; Lim, J.I.; Lee, B.R.; Mun, C.H.; Jung, Y.; Kim, S.H. Preparation of lotus-leaf-like structured blood compatible poly(ϵ-caprolactone)-block-poly(L-lactic acid) copolymer film surfaces. *Colloids Surf. B Biointerfaces* **2014**, *114*, 28–35. [CrossRef]
27. Yuan, Z.; Chen, H.; Zhang, J. Facile method to prepare lotus-leaf-like super-hydrophobic poly(vinyl chloride) film. *Appl. Surf. Sci.* **2008**, *254*, 1593–1598. [CrossRef]
28. Ensikat, H.J.; Ditsche-Kuru, P.; Neinhuis, C.; Barthlott, W. Superhydrophobicity in perfection: The outstanding properties of the lotus leaf. *Beilstein J. Nanotechnol.* **2011**, *2*, 152–161. [CrossRef]
29. Almeida, C.B.; Corradini, E.; Forato, L.A.; Fujihara, R.; Filho, J.F.L. Microstructure and thermal and functional properties of biodegradable films produced using zein. *Polímeros* **2018**, *28*, 30–37. [CrossRef]
30. Lai H.M.; Padua, G.W. Properties and microstructure of plasticized zein films. *Cereal Chem.* **1997**, *74*, 771–775. [CrossRef]
31. Escamilla-García, M.; Calderón-Domínguez, G.; Chanona-Pérez, J.; Mendoza-Madrigal, A.; Di Pierro, P.; García-Almendárez, B.E.; Amaro-Reyes, A.; Regalado-González, C.; Di Pierro, P.; García-Almendárez, B.E.; et al. Physical, structural, barrier, and antifungal characterization of chitosan-zein edible films with added essential oils. *Int. J. Mol. Sci.* **2017**, 2370. [CrossRef]
32. Muthuselvi, L.; Dhathathreyan, A. Contact angle hysteresis of liquid drops as means to measure adhesive energy of zein on solid substrates. *Pramana* **2006**, *66*, 563–574. [CrossRef]
33. Morillon, V.; Debeaufort, F.; Blond, G.; Capelle, M.; Voilley, A. Factors affecting the moisture permeability of lipid-based edible films: A review. *Crit. Rev. Food Sci. Nutr.* **2002**, *42*, 67–89. [CrossRef]
34. Shi, W.; Dumont, M.J. Review: Bio-based films from zein, keratin, pea, and rapeseed protein feedstocks. *J. Mater. Sci.* **2014**, *49*, 1915–1930. [CrossRef]

35. EAFUS—Everything Added to Food: A Food Additivie Database. Available online: http://www.cfsan.fda.gov/~dms/eafus.html (accessed on 14 September 2018).
36. SCOGS—Select Committee on Gras Substances. Available online: https://www.accessdata.fda.gov/scripts/fdcc/?set=SCOGS&sort=Sortsubstance&order=ASC&startrow=1&type=basic&search=licorice (accessed on 20 July 2018).
37. Saito, H.E.; Harp, J.R.; Fozo, E.M. *Enterococcus faecalis* Responds to Individual Exogenous Fatty Acids Independently of Their Degree of Saturation or Chain Length. *Appl. Environ. Microbiol.* **2017**, *84*, e01633-17. [CrossRef]
38. Radoshevich, L.; Cossart, P. *Listeria monocytogenes*: towards a complete picture of its physiology and pathogenesis. *Nat. Rev. Microbiol.* **2017**, *16*, 32–46. [CrossRef]
39. Avramescu, R.E.; Ghica, M.V.; Dinu-Pîrvu, C.; Prisada, R.; Popa, L. Superhydrophobic natural and artificial Surfaces-A structural approach. *Materials* **2018**, *11*. [CrossRef]

Article

Discovery of Antibacterial Dietary Spices That Target Antibiotic-Resistant Bacteria

Dan Zhang [1], Ren-You Gan [1,*], Arakkaveettil Kabeer Farha [1], Gowoon Kim [1], Qiong-Qiong Yang [1], Xian-Ming Shi [1], Chun-Lei Shi [1], Qi-Xia Luo [2], Xue-Bin Xu [3], Hua-Bin Li [4] and Harold Corke [1,*]

1 Department of Food Science & Technology, School of Agriculture and Biology, Shanghai Jiao Tong University, Shanghai 200240, China; zhang.dan@sjtu.edu.cn (D.Z.); farhatintu@sjtu.edu.cn (A.K.F.); gowoon_kim@sjtu.edu.cn (G.K.); yangqiongqiong@sjtu.edu.cn (Q.-Q.Y.); xmshi@sjtu.edu.cn (X.-M.S.); clshi@sjtu.edu.cn (C.-L.S.)

2 State Key Laboratory for Diagnosis and Treatment of Infectious Diseases, Collaborative Innovation Center for Diagnosis and Treatment of Infectious Diseases, Zhejiang University, Hangzhou 310003, China; qixia_luo@zju.edu.cn

3 Department of Microbiology, Shanghai Municipal Center for Disease Control and Prevention, Shanghai 200336, China; xxb72@sina.com

4 Guangdong Provincial Key Laboratory of Food, Nutrition and Health, Guangdong Engineering Technology Research Center of Nutrition Translation, Department of Nutrition, School of Public Health, Sun Yat-Sen University, Guangzhou 510080, China; lihuabin@mail.sysu.edu.cn

* Correspondence: renyougan@sjtu.edu.cn (R.-Y.G.); hcorke@sjtu.edu.cn (H.C.); Tel.: +86-21-3420-8517 (R.-Y.G.); +86-21-3420-8515 (H.C.)

Received: 11 April 2019; Accepted: 28 May 2019; Published: 29 May 2019

Abstract: Although spice extracts are well known to exhibit antibacterial properties, there is lack of a comprehensive evaluation of the antibacterial effect of spices against antibiotic-resistant bacteria. In the present study, ethanolic extracts from a total of 67 spices were comprehensively investigated for their in vitro antibacterial activities by agar well diffusion against two common food-borne bacteria, *Staphylococcus aureus* and *Salmonella enteritidis*, with multi-drug resistance. Results showed that *S. aureus* was generally more sensitive to spice extracts than *S. enteritidis*. Of the 67 spice extracts, 38 exhibited antibacterial activity against drug-resistant *S. aureus*, while only four samples were effective on drug-resistant *S. enteritidis*. In addition, 11 spice extracts with inhibition zones greater than 15 mm were further verified for their broad-spectrum antibacterial properties using another 10 drug-resistant *S. aureus* strains. It was found that five spice extracts, including galangal, fructus galangae, cinnamon, yellow mustard seed, and rosemary, exhibited the highest antibacterial capacity. Further cytotoxicity of these 11 spices was determined and LC_{50} values were found to be more than 100 μg/mL except for galangal, rosemary, and sage, whose LC_{50} values were 9.32 ± 0.83, 19.77 ± 2.17, and 50.54 ± 2.57, respectively. Moreover, the antioxidant activities (ferric-reducing antioxidant power (FRAP) and trolox equivalent antioxidant capacity (TEAC) values) and total phenolic content (TPC) of spice extracts were determined to establish possible correlations with the antibacterial activity. Although the antibacterial effect was positively correlated with the antioxidant activities and TPC, the correlation was weak ($r < 0.5$), indicating that the antibacterial activity could also be attributed to other components besides antioxidant polyphenols in the tested spice extracts. In conclusion, dietary spices are good natural sources of antibacterial agents to fight against antibiotic-resistant bacteria, with potential applications as natural food preservatives and natural alternatives to antibiotics in animal feeding.

Keywords: spice extracts; drug resistant bacteria; antibacterial activity; antioxidant activity; total phenolic content; correlation

1. Introduction

Food poisoning caused by food-borne bacteria is one of the critical threats to human health all over the world [1]. The emergence of multi-drug resistant bacteria induced by the abuse of antibiotics cause greater obstacles for the treatment of food-borne diseases [2]. Antibiotic-resistant bacteria, such as *Staphylococcus* (*S.*) *aureus* and *Salmonella* (*S.*) *enteritidis*, have frequently been reported to cause contamination of different foods like raw pork, beef, and poultry [3,4]. Many attempts, such as the use of synthetic preservatives, have been used to control microbial growth and ensure food safety. However, there have potential carcinogenic and toxicological properties, as well as side effects like food allergies and sensitivities that are harmful to human health [5]. To counter these problems, much effort has gone into the search for "naturally derived" alternative antimicrobials since plants are known to produce diverse secondary metabolites that are associated with anti-infective mechanisms against the invasion of pathogenic microorganisms [6,7]. Among them, plant-derived spices and extracts containing a mixture of active ingredients have received growing attention, not only for their effective antibacterial activity but also for the relative difficulty in developing resistance to them. Moreover, spice and their major components are generally recognized as safe (GRAS) with no historical records of detrimental impacts and with modern toxicological verification [8].

Although spices have been widely used in rituals, and as flavorings and coloring agents since ancient times [9], recent literature has increasingly reported on the antibacterial activity of spices against common Gram-positive and Gram-negative bacteria responsible for human infectious diseases and food safety problems [10–13]. Examples of such spices are cinnamon, oregano, nutmeg, basil, pepper, thyme, clove, rosemary, ginger, cumin, etc. However, few studies have focused on the inhibitory effects of these spices on antibiotic-resistant bacteria. The methanolic and ethanolic extracts of cinnamon, which was the most studied spice, were reported to have inhibitory effects on high level gentamicin-resistant (HLGR) enterococci, multi-drug resistant *Escherichia coli* AG100, methicillin-resistant *S. aureus* (MRSA), as well as β-lactamase producing multi-drug *Klebsiella pneumonia* and *Pseudomonas aeruginosa* [14–16]. Moreover, the antibacterial properties of spices are mostly attributed to lipophilic essential oils in most previous studies [17]. However, spices are also rich in hydrophilic antioxidants [18], such as polyphenols, many of which possess excellent antioxidant activity, and also exhibit good antibacterial activity [19]. Considering that microbial contamination and lipid oxidation are the two major factors resulting in food spoilage [20], spice hydrophilic extracts with good antibacterial and antioxidant activities can be promising natural food preservatives. For instance, extracts of cinnamon, oregano, and especially clove, were confirmed to be effective for retarding lipid oxidation and reducing pathogen numbers in real food matrices like cheese and raw pork [21,22]. More importantly, probiotic bacteria like lactic acid bacteria (LAB) were less influenced by the presence of these phenolic rich spice extracts, indicating that spice extracts could be applied in foods not only to prolong shelf-life but also enhance health benefits of foods [23].

Therefore, the aim of this study was to evaluate systemically and compare the in vitro antibacterial activity of the ethanolic extracts of 67 spices, mainly focusing on their effects on antibiotic-resistant bacteria, and to analyze the correlation among antibacterial activity, antioxidant activity, and total phenolic content (TPC) in spices. Overall, this study can shed light on the control of antibiotic-resistant bacteria using dietary spices, which should have broad applications in food industry to help assure food safety.

2. Materials and Methods

2.1. Chemicals and Reagents

2,2-azinobis (3-ethylbenzothiazoline-6-sulfonic acid) diammonium salt (ABTS) and 2,4,6-tri(2-pyridyl)-s-triazine (TPTZ) was from Sigma/Aldrich (St. Louis, MO, USA). Gallic acid was from Energy Chemical (Shanghai, China). Dimethyl sulfoxide (DMSO) was from Beyotime (Shanghai, China), Folin–Ciocalteu reagent was from Macklin (Shanghai, China),

6-hydroxy-2,5,7,8-tetramethylchromane-2-carboxylic acid (trolox) was from Fluka Chemika AG (Buchs, Switzerland). Hydrochloric acid, iron (III) chloride hexahydrate and Iron (II) sulfate heptahydrate were from Sinopharm Chemical Reagent (Shanghai, China), acetic acid and sodium acetate were from Molbase (Shanghai, China). Potassium persulphate and ethanol were obtained from Titanchem (Shanghai, China). Sodium carbonate and methanol were purchased from J&K (Beijing, China). Luria Bertani (LB) broth, agar bacteriological, and Mueller–Hinton (MH) broth were purchased from Oxiod (Basingstoke, England). Antibiotics, including ampicillin, cefazolin, ciprofloxacin, clindamycin, erythromycin, gentamicin, oxacillin, penicillin, streptomycin, sulfisoxazole, and tetracycline were purchased from Meilune (Dalian, China). All chemicals used in the experiment were of analytical grade.

2.2. Spice Materials

The 67 dried edible spice materials were purchased from the local markets in Shanghai, China. The basic information (scientific name and common name) of these spices is detailed in Table 1.

2.3. Microorganisms and Culture

Two strains of *Salmonella enteritidis* (drug-resistant *S. enteritidis* SJTUF 10987 and the standard strain *S. enteritidis* ATCC 13076) and 12 strains of *Staphylococcus aureus*, including 11 drug-resistant strains *S. aureus* SJTUF 20745, *S. aureus* SJTUF 20746, *S. aureus* SJTUF 20755, *S. aureus* SJTUF 20758, *S. aureus* SJTUF 20772, *S. aureus* SJTUF 20827, *S. aureus* SJTUF 20841, *S. aureus* SJTUF 20862, *S. aureus* SJTUF 20973, *S. aureus* SJTUF 20978, and *S. aureus* SJTUF 20991, as well as the standard strain *S. aureus* ATCC 25923 were used in this study. These strains were stored at −80 °C. To prepare the inocula, a single colony of the bacteria grown on the LB agar plate was selected and transferred into the LB broth to culture overnight in a rotary incubator (37 °C, 150 rpm). The bacterial suspension was then diluted to approximately 1×10^6 colony-forming units (CFU)/mL for subsequent antibacterial experiments.

Table 1. Antibacterial properties of 67 spice extracts.

Scientific Name	Common Name	Parts Tested	Diameters of Inhibitory Zone (DIZ, mm)			
			S. aureus SJTUF 20978 (Resistant)	*S. aureus* ATCC 25923 (Normal)	*S. enteritidis* SJTUF 10987 (Resistant)	*S. enteritidis* ATCC 13076 (Normal)
Alpinia galangal (L.) Willd.	Galangal	Rhizome	25.6 ± 0.49	31.7 ± 0.21	NIZ	NIZ
Alpinia galanga Willd.	Fructus galangae	Fruit	20.2 ± 0.52	28.3 ± 0.29	NIZ	NIZ
Alpinia hainanensis K. Schum.	Semen alpiniae katsumadai	Fruit	14.8 ± 0.74	17.5 ± 0.09	NIZ	NIZ
Alpinia officinarum Hance	Small galangal	Rhizome	11.4 ± 0.24	11.9 ± 0.14	NIZ	NIZ
Alpinia tonkinensis Gagnep	Green gardamon	Fruit	NIZ	NIZ	NIZ	NIZ
Amomum aurantiacum H. T. Tsai et S. W. Zhao	Thorn amomum villosum	Fruit	13.4 ± 0.09	15.3 ± 0.24	8.6 ± 0.21	9.1 ± 0.24
Amomum testaceum Ridl	Fructus amomi rotundus	Fruit	12.0 ± 0.29	13.5 ± 0.29	NIZ	NIZ
Amomum tsao-ko Crevost et Lemarié	Fructus tsaoko	Fruit	14.1 ± 0.09	14.8 ± 0.24	NIZ	NIZ
Amomum villosum Lour.	Fructus amomi	Fruit	12.1 ± 0.24	14.9 ± 0.09	NIZ	NIZ
Anethum graveolens L.	Dill	Seed	NIZ	NIZ	NIZ	NIZ
Angelica dahurica (Hoffm.) Benth. et Hook.f. ex Franch. et Sav.	Radix angelicae formosanae	Rhizome	NIZ	NIZ	NIZ	NIZ
Areca catechu L.	Areca	Fruit	10.1 ± 0.09	9.90 ± 0.09	NIZ	NIZ
Artemisia dracunculus L.	Tarragon	Leaf	NIZ	NIZ	NIZ	NIZ
Aucklandia lappa Decne.	Costustoot	Rhizome	10.1 ± 0.19	15.2 ± 0.29	NIZ	NIZ
Capsicum annuum L.	Dry chilli (grown in Henan)	Fruit	NIZ	NIZ	NIZ	NIZ
Capsicum annuum L.	Dry chilli (grown in Sichuan)	Fruit	NIZ	NIZ	NIZ	NIZ
Capsicum annuum L.	Dry chilli (grown in Yunnan)	Fruit	NIZ	NIZ	NIZ	NIZ
Capsicum annuum var. grossum	Bell pepper	Fruit	NIZ	NIZ	NIZ	NIZ
Carum carvi L.	Caraway	Fruit	8.70 ± 0.09	10.6 ± 0.09	NIZ	NIZ
Cinnamomum cassia (L.) J.Presl	Cinnamon	Bark	20.7 ± 0.47	27.6 ± 1.73	16.0 ± 0.52	15.5 ± 0.38
Citrus limon (L.) Osbeck	Dried lemon	Fruit	11.3 ± 0.09	13.0 ± 0.09	NIZ	NIZ
Citrus reticulata Blanco	Citrus	Fruit	NIZ	NIZ	NIZ	NIZ
Citrus reticulata Blanco	Old citrus	Fruit	NIZ	NIZ	NIZ	NIZ
Coriandrum sativum L	Coriander	Fruit	NIZ	NIZ	NIZ	NIZ
Crataegus pinnatifida Bunge	Hawthorn	Fruit	11.4 ± 0.50	12.1 ± 0.29	NIZ	NIZ
Cuminum cyminum L.	Chinese cumin seed	Fruit	NIZ	NIZ	NIZ	NIZ
Curcuma longa L.	Turmeric	Rhizome	NIZ	NIZ	NIZ	NIZ
Cymbopogon citratus (DC.) Stapf.	Lemongrass	Leaf	NIZ	NIZ	NIZ	NIZ
Eleutherococcus nodiflorus (Dunn) S.Y.Hu.	Cortex acanthopanacis	Bark	NIZ	NIZ	NIZ	NIZ
Foeniculum vulgare Mill.	Fennel (traditional Chinese spice)	Fruit	NIZ	NIZ	NIZ	NIZ
Foeniculum vulgare	Kelly anise seeds (Western food spice)	Fruit	NIZ	NIZ	NIZ	NIZ
Gardenia jasminoides J. Ellis	Gardenia	Fruit	NIZ	NIZ	NIZ	NIZ
Glycyrrhiza uralensis Fisch.	Liquorice	Leaf	15.8 ± 0.14	15.6 ± 0.57	NIZ	NIZ
Illicium verum Hook. f.	Star anise	Fruit	12.5 ± 0.09	12.8 ± 0.29	NIZ	NIZ
Kaempferia galanga L.	Rhizoma kaempferiae	Rhizome	10.1 ± 0.49	10.3 ± 0.52	NIZ	NIZ
Laurus nobilis L.	Bay leaf	Leaf	12.5 ± 0.33	13.0 ± 0.09	NIZ	NIZ
Lithospermum erythrorhizon Sieb. et Zucc.	Lithospermum	Leaf	13.3 ± 0.21	13.3 ± 0.28	NIZ	9.6 ± 0.29
Lysimachia capillipes Hemsl	Nephrolepis	Stem	12.4 ± 0.45	13.9 ± 0.62	NIZ	NIZ
Lysimachia foenum-graecum Hance	Avandula pedunculata	Whole plant	12.3 ± 0.49	14.4 ± 0.33	NIZ	10.9 ± 0.24
Magnolia denudata Desr.	Magnolia flower	Flower	NIZ	NIZ	NIZ	NIZ
Mentha canadensis L.	Pepper mint	Leaf	12.5 ± 0.09	15.9 ± 0.33	NIZ	NIZ
Monascus purpureus Went	Red yeast rice	Fruit	NIZ	NIZ	NIZ	NIZ
Murraya koenigii (L.) Spreng.	Curry leaves	Leaf	9.80 ± 0.61	10.8 ± 0.09	NIZ	NIZ
Murraya paniculata (L.) Jack.	Murraya paniculata	Leaf	NIZ	NIZ	NIZ	NIZ
Myristica fragrans Houtt.	Semen myristicae	Fruit	10.5 ± 0.21	11.6 ± 0.29	NIZ	NIZ
Nardostachys jatamansi (D. Don) DC.	Nard	Stem	14.1 ± 0.47	15.5 ± 0.21	NIZ	NIZ
Ocimum basilicum L.	Basil	Leaf	NIZ	NIZ	NIZ	NIZ

Table 1. *Cont.*

Scientific Name	Common Name	Parts Tested	Diameters of Inhibitory Zone (DIZ, mm)			
			S. aureus SJTUF 20978 (Resistant)	*S. aureus* ATCC 25923 (Normal)	*S. enteritidis* SJTUF 10987 (Resistant)	*S. enteritidis* ATCC 13076 (Normal)
Origanum majorana L.	Marjoram	Whole plant	17.2 ± 0.71	15.2 ± 0.62	NIZ	NIZ
Origanum vulgare L.	Origanum	Leaf	14.1 ± 0.39	11.8 ± 0.09	NIZ	NIZ
Petroselinum crispum (Mill.) Fuss	Parsley	Leaf	NIZ	NIZ	NIZ	NIZ
Pimenta dioica (L.) Merr.	Allspice	Fruit	13.2 ± 0.09	14.9 ± 0.19	NIZ	NIZ
Piper longum L.	Long pepper	Cluster	NIZ	NIZ	NIZ	NIZ
Piper nigrum L.	Black pepper	Fruit	NIZ	NIZ	NIZ	NIZ
Piper nigrum L.	White pepper	Fruit	NIZ	NIZ	NIZ	NIZ
Piper nigrum L.	Red pepper	Fruit	14.2 ± 0.14	16.8 ± 0.09	NIZ	NIZ
Piper nigrum L.	Green pepper	Fruit	NIZ	NIZ	NIZ	NIZ
Reseda odorata L.	Integrated vanilla	Leaf	14.7 ± 0.51	14.0 ± 0.75	NIZ	NIZ
Rosmarinus officinalis L.	Rosemary	Leaf	18.3 ± 0.21	20.8 ± 0.14	NIZ	10.3 ± 0.42
Salvia japonica Thunb.	Sage	Leaf	15.4 ± 0.49	19.3 ± 0.21	NIZ	NIZ
Sinapis alba L.	Yellow mustard seeds	Seed	18.6 ± 1.03	21.9 ± 0.16	NIZ	NIZ
Sinapis alba L.	Black mustard seeds	Seed	NIZ	NIZ	NIZ	NIZ
Sophora alopecuroides L.	Fenugreek	Fruit	NIZ	NIZ	NIZ	NIZ
Syzygium aromaticum (L.) Merr. et L. M. Perry	Male clove	Flower	15.6 ± 0.33	16.4 ± 0.39	11.0 ± 0.21	9.0 ± 0.29
Syzygium aromaticum	Female clove	Fruit	15.1 ± 0.49	20.9 ± 0.14	8.5 ± 0.24	10.6 ± 0.09
Thymus vulgaris L.	Thyme	Leaf	16.0 ± 0.33	14.9 ± 0.42	NIZ	NIZ
Zanthoxylum bungeanum Maxim	Red Chinese prickly ash	Fruit	11.3 ± 0.09	12.9 ± 0.21	NIZ	NIZ
Zanthoxylum bungeanum Maxim	Green Chinese prickly ash	Fruit	13.3 ± 0.24	13.3 ± 0.21	NIZ	NIZ

The data were averages of three measurements with standard deviation. NIZ means no inhibition zone.

2.4. Verification of the Drug-Resistant Bacteria

The verification assay was undertaken using the agar dilution method following the testing procedures of the Clinical and Laboratory Standards Institute, USA (CLSI M100-S28, 2018). Briefly, tested strains were cultured in MH broth overnight, then centrifuged (1000× *g*, 1 min) and transferred to a custom 96-well microtitre plate using 1 mL of 0.85% sodium chloride solution. The absorbance of bacterial resuspension at OD 600 was adjusted to 0.5, followed by further 100 fold dilution to make a final concentration of 1×10^6 CFU/mL. Subsequently, bacterial suspensions were incubated to the MH agar plates which were supplemented with antibiotics under prescribed breakpoint concentrations stated by CLSI using the multipoint incubator (HM1-12-001, Hen Gao Technology Development Co., Ltd., Tianjin, China), and the plates were cultured at 37 °C for 20 h. In addition, the oxacillin agar dilution method was used for the detection of MRSA (CLSI, 2018). Isolates were considered to be multi-drug resistant if they were resistant to at least three different categories of antibiotics, based on their growth condition on the MH plates.

2.5. Preparation of Spice Ethanolic Extracts

The clean dietary spices were air-dried in a ventilated oven at 40 °C for 24 h, then ground into fine powders by a miller (S025, IKA, Staufen, Germany). Powdered samples (4.0 g) were extracted with 80 mL of 80% (*v*/*v*) ethanol in a shaking bath (MQT-50, Shanghai Min Quan Co., Ltd., Shanghai, China) at room temperature (23 ± 1 °C) for 24 h. In this study, 80% ethanol was used for the extraction, since ethanol is of relative low toxicity among several organic solvents, and 80% ethanol was efficient to extract antioxidant and antibacterial components from spices and herbs based on our previous study [22]. Afterwards, the mixture was centrifuged at room temperature (900× *g*, 15 min) and the collected supernatant was concentrated by a rotatory vacuum evaporator (RE-52AA, Shanghai Ya Rong Co., Ltd., Shanghai, China) at 40 °C, then the concentrated extract was dried by a vacuum freeze-dryer (SJIA-5FE, Ningbo Shuang Jia instrument Co., Ltd., Ningbo, China). The freeze-dried samples were stored at −20 °C in small vials for further use.

2.6. Determination of Antibacterial Activity

The inhibitory effects of 67 spices were estimated according to the agar diffusion method as previously reported with slight modification [24]. The freeze-dried extracts were dissolved in DMSO to a final concentration of 100 mg/mL and filtered through 0.22 μm sterilizing filters. Briefly, all bacteria were diluted to about 1×10^6 CFU/mL with sterile LB medium, and then 100 μL of each bacterial suspension was evenly spread onto the surface of LB agar plate by sterile glass beads (6 mm in diameter). Oxford cups (sterilized hollow cylinder with an inner diameter of 6 mm, outer diameter of 7.8 mm, and height of 10 mm) were placed lightly on the agar surface, and then 60 μL of the prepared samples (100 mg/mL) were delivered into the cups. DMSO (60 μL/cup) was used as a negative control. The plates were incubated at 37 °C for 24 h for bacterial growth in an incubator (BI-150A, Shanghai Stik Co., Ltd., Shanghai, China). The diameters of the inhibitory zones (DIZs) formed around the Oxford cups were measured to evaluate the antibacterial activity and expressed in millimeter (mm). All experiments were performed in triplicate. DIZ values less than 8 mm were considered as "no inhibition zone (NIZ)".

2.7. Minimum Inhibitory Concentration (MIC) and Minimum Bactericide Concentration (MBC) Assays

Minimum inhibitory concentration (MIC) and minimum bactericide concentration (MBC) were determined according to the method described by Elshikh et al. with minor adjustment [25]. Briefly, 100 μL of MH broths were added into 96-well plate and then another 100 μL of dissolved samples, whose initial concentrations were 100 mg/mL, were added in first wells. Serial two-fold dilutions were made with final concentrations ranging from 50 mg/mL to 0.39 mg/mL. Afterwards, 100 μL of the standardized bacteria suspensions (1×10^6 CFU/mL) was added to each test well, so that the final volume in each well was 200 μL. Plates were incubated at 37 °C for 24 h. After incubation, 30 μL of the freshly prepared resazurin (0.015%) was added to all test wells, and further incubated for 2 h to allow the viable microorganism to metabolize the blue resazurin dye into pink resorufin. MIC was defined as the concentration at which the corresponding well showed no color change. Afterwards, the contents of wells with concentrations equal or higher than MIC values were directly incubated onto the MH plate, and the lowest concentration at which there was no colony growth was defined as MBC.

2.8. Cytotoxicity of Spice Extracts

Human foreskin fibroblast (HFF) cells were grown in DMEM medium supplemented with 10% fetal bovine serum (FBS), and maintained at 37 °C in a humidified 5% CO_2 incubator. The toxicity of spice extracts was evaluated by 3-(4,5)-dimethylthiazol-2-yl)-2,5-diphenyltetrazolium bromide (MTT) assay as described by Mosmann with slight modification [26]. HFF cells were seeded at a density of 5×10^4 cells/mL into the 96-well plate and incubated overnight to adhere the cells. The cells were treated with various concentrations (5–100 μg/mL, and the concentration of DMSO was diluted below 0.1%) of different extracts and incubated at 37 °C in a humidified 5% CO_2 incubator for 24 h. The untreated cells were included as control. After the incubation period, MTT (20 μL of 5 mg/mL) was added into each well and incubated for 4 h. The formazan crystals were dissolved with DMSO (100 μL). The absorbance was measured at 570 nm using a microtitre plate reader (SpectraMax iD3, Molecular Devices, Silicon Valley, NC, USA). The LC_{50} value was calculated as the concentration of the extract that resulted in 50% reduction of absorbance compared to control cells [27].

2.9. Determination of Antioxidant Capacity

2.9.1. Ferric-Reducing Antioxidant Power (FRAP) Assay

The ferric-reducing antioxidant power (FRAP) assay was carried out according to the procedures described by Gan et al. [28]. Briefly, the FRAP working solution was freshly prepared before the experiment, with sodium acetate solution (300 mM, pH = 3.7), TPTZ (10 mM solved with HCl) and ferric chloride solution (20 mM) mixedin a volume ratio of 10:1:1, respectively. The FRAP working solution was then incubated at 37 °C before use. The proper dilutions (100 μL) of samples were added to 3 mL of the FRAP working solution and their absorbance at 593 nm was determined after incubation for 4 min at room temperature (23 ± 1 °C). Ferrous sulfate solution (0.1–1 mM) was used as the standard for the calibration curve, and the results were expressed as mmol Fe(II)/g dry weigh (DW) extract powder. All tests were performed in triplicate.

2.9.2. Trolox Equivalent Antioxidant Capacity (TEAC) Assay

The trolox equivalent antioxidant capacity (TEAC) assay was carried out to determine the free radical scavenging capacity using $ABTS^+$ according to the method previously reported [28]. ABTS stock solution was prepared by mixing 7 mM ABTS and 2.45 mM potassium persulfate in a ratio of 1:1 (*v*/*v*) and then incubated at room temperature (23 ± 1 °C) in the dark for at least 16 h. The ABTS working solution was prepared by dilution of the stock solution with 80% ethanol before use, and then the absorbance at a wavelength of 734 nm was adjusted to 0.7 ± 0.05. The blank control is a mixture of 3.9 mL ABTS working solution and 0.1 mL of 80% ethanol. The spice extract sample (0.1 mL) was diluted with 80% ethanol to provide 20–80% inhibition of the blank absorbance, and then the properly diluted samples were added to 3.9 mL ABTS working solutions and mixed thoroughly. The absorbance of reactive mixture was determined at 734 nm after incubation at room temperature (23 ± 1 °C) for 6 min. Quantitative results were determined from the standard curve of trolox (0.05–0.8 mM) and were expressed as mmol trolox/g dry weight (DW) extract powder. All tests were performed in triplicate.

2.10. Determination of Total Phenolic Content (TPC)

TPC was determined by the Folin-Ciocalteu method reported previously with some modification [29]. The appropriate dilutions of samples (200 μL) were mixed with 1 mL of 0.5 M Folin–Ciocalteu reagent at room temperature (23 ± 1 °C) for 4 min, and then reacted with 800 μL of saturated sodium carbonate solution (75 g/L) in dark for 2 h. Finally, the absorbance of the reaction mixtures was measured at 760 nm with a spectrophotometer (181712007PC, Shanghai Jing Hua Co. Ltd., Shanghai, China) and quantified on the base of the standard curve of gallic acid (0.01–0.1 mM). The results were expressed as milligram gallic acid equivalent (mg GAE)/g DW extract powder. All tests were performed in triplicate.

2.11. Statistical Analysis

All the measurements were performed in triplicate, and the results were expressed as mean ± standard deviation (SD). Statistical analysis was performed using Microsoft Excel 2016 (Microsoft, Seattle, MA, USA) and SPSS 22.0 (IBM SPSS Statistics, IBM Corp, Somers, NY, USA). Pearson linear correlation analysis and principal component analysis (PCA) were performed to analyze relationships among parameters, and p value less than 0.01 was defined as statistical significance.

3. Results

3.1. Verification of Drug-Resistant Bacteria

The antibiotic-resistant strains of *S. aureus* and *S. enteritidis* were isolated from food samples in our lab previously. In order to confirm their antibiotic resistance, we first tested their resistance to 11 common antibiotics from different categories. The breakpoint concentration of each antibiotic defined by CLSI and corresponding bacterial resistance spectra are shown in Table 2. All tested bacteria were resistant to antibiotics. Except for *S. aureus* SJTUF 20827 and *S. aureus* SJTUF 20973, the remaining bacteria were identified as multi-drug resistant bacteria, resistant to at least three antibiotics. These bacteria showed the highest resistance rate to erythromycin, reaching up to 83.3%, followed by ciprofloxacin (75%), Clindamycin (75%), gentamicin (50%), and streptomycin (50%). However, no strain resistant to oxacillin (methicillin-resistant *Staphylococcus aureus*, MRSA) was detected. Overall, it was confirmed that all selected bacteria were antibiotic-resistant, most of which were multi-drug resistant.

Table 2. Verification of drug-resistant bacteria.

Name of Antibiotics	Class	Breakpoint Conc. (μg/mL)	*S. enteritidis* SJTUF 10987	*S. aureus* SJTUF										
				20745	20746	20755	20758	20772	20827	20841	20862	20973	20978	20991
Ampicillin	β-lactams	32	+											
Cefazolin		8												
Oxacillin		4												
Penicillin		0.25					+							
Gentamicin	Aminoglycosides	16			+			+		+	+	+		
Streptomycin		64	+	+		+	+						+	
Ciprofloxacin	Fluoroquinolones	4	+	+	+	+		+		+	+	+		+
Clindamycin	Lincosamides	4		+	+	+	+	+		+	+		+	+
Erythromycin	Macrolides	8		+	+	+	+	+	+	+	+		+	+
Sulfisoxazole	Sulfonamides	512	+											
Tetracycline	Tetracyclines	16												+

+ means bacterial growth.

3.2. Antibacterial Activity against Antibiotic-Resistant Bacteria

The agar diffusion method was used to evaluate systematically the antibacterial activity of 67 spice extracts on antibiotic-resistant *S. aureus* SJTUF 20978 and *S. enteritidis* SJTUF 10987, with each standard strain used as the comparison. The results of DIZ were presented in Table 1. A significant variation in the antibacterial activity reflected by different DIZ values was observed, depending on the type of spice extracts and the subject bacteria.

For antibiotic-resistant *S. aureus* SJTUF 20978, a significant proportion of the spices (38, approximately accounting for 56.7% of the total tested samples) exhibited antibacterial activity, with DIZ in the range of 8.7–25.6 mm. Besides, 11 of these spice extracts (accounting for 16.4%) showed a relatively superior antibacterial activity, with DIZ values greater than 15 mm. Among there, galangal, the rhizome of *Alpinia galanga* (L.) Willd., showed exceptional antibacterial capacity, with the DIZ reaching 25.6 mm, followed by cinnamon, fructus galangae (ripe fruit from galangal), yellow mustard seed, rosemary, and marjoram, with DIZ values of 20.7, 20.2, 18.6, 18.3, and 17.2 mm, respectively. Moreover, the MIC and MBC results of these 11 samples were determined and were shown in Table 3. The MIC values ranged from 0.40–6.25 mg/mL, and MBC values ranged from 0.40–12.5 mg/mL, which were one or two times higher than MICs. Among them, clove fruit, sage, rosemary, and liquorice had the lowest MIC value (0.40 mg/mL), and fructus galangae, galangal, and yellow mustard seed showed a relatively high value (6.25 mg/mL). However, for drug-resistant *S. enteritidis* SJTUF 10987, the overall antibacterial effects of tested spice extracts were relatively low, and only four samples showed inhibitory activity, including cinnamon (DIZ = 16.0 mm), male clove (DIZ = 11.0 mm), thorn amomum villosum (DIZ = 8.6 mm), and female clove (DIZ = 8.5 mm).

Table 3. Minimum inhibitory concentration (MIC) and minimum bactericide concentration (MBC) values of selected 11 spice extracts with good antibacterial activity.

Scientific Name	Common Name	MIC (mg/mL)	MBC (mg/mL)
Alpinia galangal (L.) Willd.	Galangal	6.25	6.25
Alpinia galanga Willd.	Fructus galangae	6.25	6.25
Cinnamomum cassia (L.) J. Presl	Cinnamon	0.8	1.6
Glycyrrhiza uralensis Fisch.	Liquorice	0.4	0.8
Origanum majorana L.	Marjoram	1.6	1.6
Rosmarinus officinalis L.	Rosemary	0.4	0.4
Salvia japonica Thunb.	Sage	0.4	0.8
Sinapis alba L.	Yellow mustard seeds	6.25	12.5
Syzygium aromaticum (L.) Merr. et L. M. Perry	Male clove (flower)	0.8	1.6
Syzygium aromaticum (L.) Merr. et L. M. Perry	Female clove (fruit)	0.4	0.4
Thymus vulgaris L.	Thyme	1.6	1.6

In general, the antibacterial activity of spice extracts against antibiotic-resistant bacteria was somewhat less effective compared to corresponding standard strains *S. aureus* ATCC 25923 and *S. enteritidis* ATCC 13076 (data shown in Table 1). Besides, spice extracts showed much better antibacterial activity against Gram-positive *S. aureus* than Gram-negative *S. enteritidis*. Therefore, we further tested whether spice extracts had a broad spectrum antibacterial effect on drug-resistant *S. aureus*. 11 spice extracts with DIZ more than 15 mm on drug-resistant *S. aureus* SJTUF 20978 were selected to verify their antibacterial capacity against another 10 antibiotic-resistant strains of *S. aureus*. As shown in Figure 1, all selected spice extracts exhibited inhibitory effects against the validated antibiotic-resistant strains of *S. aureus*. Among them, galangal (DIZ = 25.6–31.1 mm), fructus galangae (DIZ = 21.2–27.8 mm), and cinnamon (DIZ = 18.3–25.1 mm) showed the strongest antibacterial effects, followed by yellow mustard seed (DIZ = 18.0–21.4 mm) and rosemary (DIZ = 16.2–19.9 mm). Overall, selected spice extracts possessed a broad spectrum antibacterial effect against antibiotic-resistant *S. aureus*.

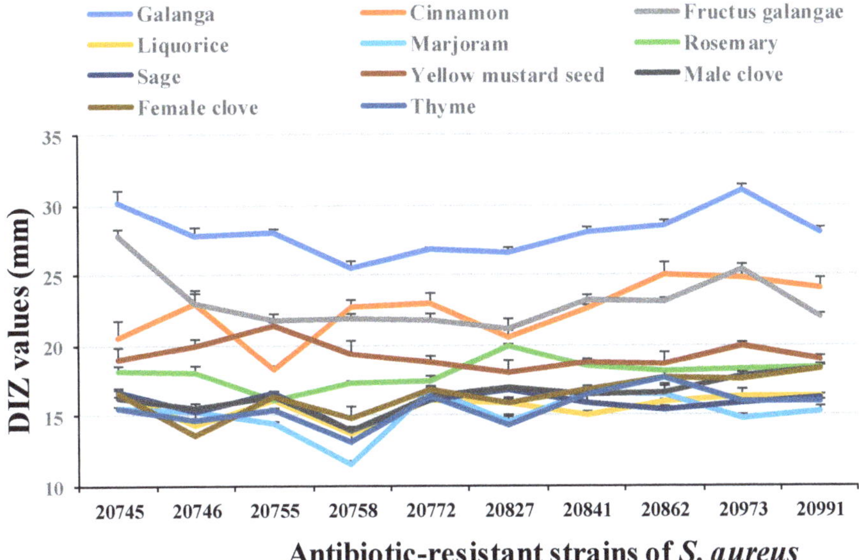

Figure 1. The antibacterial activity of selected spice extracts on 10 antibiotic-resistant strains of *S. aureus*.

3.3. Cytotoxicity of Spice Extracts

Spices were generally considered to be non-toxic or less toxic because of their natural origin and long use as food additives and medicine for ailment treatments. Meanwhile, some studies on the efficacy and safety of plants pointed out that some phytochemicals displayed certain cytotoxicity, genotoxicity, and carcinogenic effects when used chronically [30]. Therefore, it was necessary to determine the cytotoxicity of the selected 11 spice extracts with good activity against multi-drug resistant *S. aureus*. The cytotoxicity of these chosen 11 spices were determined by using the in vitro assay with HFF cells, and LC_{50} values were calculated. Of 11 spices tested, eight spice extracts did not show any cytotoxicity against HFF cells after 24 h of treatment with the highest concentration tested (100 μg/mL), suggesting that their LC_{50} values higher than 100 μg/mL. However, other three spices, including galangal, rosemary, and sage, were able to inhibit the growth of HFF cells at the LC_{50} of 9.32 ± 0.83, 19.77 ± 2.17, and 50.54 ± 2.57 μg/mL, respectively, indicating their potential safety issue. Overall, the results found that most spice extracts with high antibacterial effect were low toxic, and could be used as potential antimicrobial agents in food industry.

3.4. Antioxidant Activity of Spice Extracts

The antioxidant activity of spice extracts was determined using FRAP and TEAC assays, and the results are shown in Table 4. The strongest antioxidant activity determined by FRAP assay was found in the extract of female clove, followed by male clove, allspice, red pepper, and fructus amomi, showing the FRAP values about 6682 ± 68.6, 5453 ± 23.9, 4404 ± 23.9, 4137 ± 147 and 3605 ± 201 mmol Fe (II)/g DW extract powder. In addition, the antioxidant activity measured by TEAC method was also highest in the extract of female clove, followed by male clove, semen alpiniae katsumadai, allspice, and fructus amomi. The TEAC values were 3415 ± 53.1, 3131 ± 177, 2662 ± 83.7, 2184 ± 43.9, and 2153 ± 370 mmol Trolox/g DW extract powder, respectively. Besides, the lowest antioxidant activity was found in the extract of dried lemon, with FRAP and TEAC values about 50.3 mmol Fe (II)/g DW extract powder and 17.4 mmol Fe (II)/g DW extract powder, respectively.

Table 4. Antioxidant activity and total phenolic content of 67 spice extracts.

Scientific Name	Common Name	TPC (mg GAE/g DW)	FRAP (mmol Fe (II)/g DW)	TEAC (mmol Trolox/g DW)
Alpinia galangal (L.) Willd.	Galangal	119 ± 6.41	608 ± 55.6	394 ± 27.1
Alpinia galanga Willd.	Fructus galangae	122 ± 2.49	687 ± 30.8	423 ± 58.0
Alpinia hainanensis K. Schum.	Semen alpiniae katsumadai	473 ± 8.67	2876 ± 197	2662 ± 83.7
Alpinia officinarum Hance	Small galangal	281 ± 15.3	967 ± 43.5	707 ± 47.5
Alpinia tonkinensis Gagnep	Green gardamon	64.9 ± 4.72	617 ± 24.0	284 ± 19.5
Amomum aurantiacum H. T. Tsai et S. W. Zhao	Thorn amomum villosum	350 ± 11.3	3433 ± 137	1836 ± 127
Amomum testaceum Ridl	Fructus amomi rotundus	84.7 ± 2.67	1220 ± 99.6	288 ± 25.4
Amomum tsao-ko Crevost et Lemarié	Fructus tsaoko	303 ± 0.78	2542 ± 184	1902 ± 123
Amomum villosum Lour.	Fructus amomi	360 ± 6.80	3605 ± 201	2153 ± 370
Anethum graveolens L.	Dill	115 ± 1.83	867 ± 51.2	234 ± 11.3
Angelica dahurica (Hoffm.) Benth. et Hook.f. ex Franch. et Sav.	Radix angelicae formosanae	16.6 ± 0.48	208 ± 20.9	75.3 ± 8.74
Areca catechu L.	Areca seed	95.9 ± 3.71	741 ± 35.8	390 ± 41.9
Artemisia dracunculus L.	Tarragon leaf	148 ± 5.40	1118 ± 3.15	461 ± 33.7
Aucklandia lappa Decne.	Costustoot	21.7 ± 1.90	288 ± 7.51	105 ± 3.65
Capsicum annuum L.	Dry chilli (grown in Henan)	28.4 ± 1.55	160 ± 7.16	99.2 ± 16.8
Capsicum annuum L.	Dry chilli (grown in Sichuan)	17.3 ± 0.14	142 ± 4.26	59.6 ± 5.20
Capsicum annuum L.	Dry chilli (grown in Yunnan)	29.2 ± 1.59	270 ± 5.45	109 ± 4.98
Capsicum annuum var. grossum	Bell pepper	16.1 ± 3.18	105 ± 12.7	77.2 ± 7.52
Carum carvi L.	Caraway	42.3 ± 1.62	369 ± 33.3	165 ± 54.0
Cinnamomum cassia (L.) J.Presl	Cinnamon	349 ± 12.0	3013 ± 99.0	1857 ± 47.9
Citrus limon (L.) Osbeck	Dried lemon	7.35 ± 0.68	50.3 ± 2.32	17.4 ± 0.62
Citrus reticulata Blanco	Citrus	39.7 ± 1.10	233 ± 11.4	241 ± 9.14
Citrus reticulata Blanco	Old citrus	68.2 ± 2.40	306 ± 13.8	218 ± 12.3
Coriandrum sativum L	Coriander	31.4 ± 0.42	306 ± 2.84	111 ± 9.41
Crataegus pinnatifida Bunge	Hawthorn	98.9 ± 2.09	768 ± 46.4	414 ± 10.1
Cuminum cyminum L.	Chinese cumin seed	58.8 ± 2.08	465 ± 18.6	184 ± 8.84
Curcuma longa L.	Turmeric	251 ± 4.30	1444 ± 51.5	1489 ± 252
Cymbopogon citratus (DC.) Stapf.	Lemongrass	153 ± 1.71	1586 ± 96.9	510 ± 34.5
Eleutherococcus nodiflorus (Dunn) S.Y.Hu.	Cortex acanthopanacis	79.1 ± 3.64	704 ± 55.1	372 ± 5.41
Foeniculum vulgare Mill.	Fennel (traditional Chinese spice)	58.1 ± 2.36	365 ± 16.4	222 ± 27.9
Foeniculum vulgare	Kelly anise seeds (Western spice)	30.5 ± 2.99	356 ± 8.33	221 ± 36.9
Gardenia jasminoides J. Ellis	Gardenia	45.4 ± 3.92	553 ± 17.4	152 ± 8.22
Glycyrrhiza uralensis Fisch.	Liquorice	65.8 ± 3.08	363 ± 13.0	365 ± 23.7
Illicium verum Hook. f.	Star anise	165 ± 5.00	1650 ± 35.5	855 ± 43.0
Kaempferia galanga L.	Rhizoma kaempferiae	15.9 ± 0.14	65.3 ± 2.73	23.0 ± 0.67
Laurus nobilis L.	Bay leaf	182 ± 1.46	1255 ± 81.9	1124 ± 100
Lithospermum erythrorhizon Sieb. et Zucc.	Lithospermum	80.9 ± 5.89	627 ± 6.44	325 ± 12.9
Lysimachia capillipes Hemsl	Nephrolepis	79.1 ± 0.91	709 ± 39.4	306 ± 32.2
Lysimachia foenum-graecum Hance	Avandula pedunculata	98.5 ± 8.57	759 ± 45.7	366 ± 32.8
Magnolia denudata Desr.	Magnolia flower	63.2 ± 3.71	612 ± 18.6	244 ± 10.1
Mentha canadensis L.	Pepper mint	280 ± 2.97	3180 ± 167	1296 ± 29.5
Monascus purpureus Went	Red yeast rice	65.8 ± 2.32	224 ± 5.94	158 ± 36.7
Murraya koenigii (L.) Spreng.	Curry leaves	146 ± 5.43	463 ± 27.9	241 ± 8.47

Table 4. *Cont.*

Scientific Name	Common Name	TPC (mg GAE/g DW)	FRAP (mmol Fe (II)/g DW)	TEAC (mmol Trolox/g DW)
Murraya paniculata (L.) Jack.	Murraya paniculata	70.4 ± 1.09	497 ± 1.36	349 ± 86.9
Myristica fragrans Houtt.	Semen myristicae	111 ± 3.66	934 ± 27.3	552 ± 21.7
Nardostachys jatamansi (D. Don) DC.	Nard	103 ± 2.08	553 ± 29.4	219 ± 29.0
Ocimum basilicum L.	Basil	147 ± 4.71	1410 ± 24.6	487 ± 12.3
Origanum majorana L.	Marjoram	303 ± 3.92	3272 ± 130	1563 ± 125
Origanum vulgare L.	Origanum	204 ± 0.69	2164 ± 135	714 ± 21.0
Petroselinum crispum (Mill.) Fuss	Parsley	58.4 ± 1.89	521 ± 24.1	245 ± 34.8
Pimenta dioica (L.) Merr.	Allspice	339 ± 1.36	4404 ± 23.9	2184 ± 43.9
Piper longum L.	Long pepper	92.0 ± 1.20	1733 ± 68.9	789 ± 44.2
Piper nigrum L.	Black pepper	64.3 ± 7.74	576 ± 13.0	239 ± 9.82
Piper nigrum L.	White pepper	36.8 ± 0.96	354 ± 9.54	251 ± 50.7
Piper nigrum L.	Red pepper	378 ± 3.52	4137 ± 147	2055 ± 76.4
Piper nigrum L.	Green pepper	73.5 ± 3.91	699 ± 33.2	262 ± 9.39
Reseda odorata L.	Integrated vanilla	269 ± 5.74	2671 ± 120	801 ± 42.9
Rosmarinus officinalis L.	Rosemary	261 ± 9.29	2712 ± 160	821 ± 20.8
Salvia japonica Thunb.	Sage	204 ± 3.17	2345 ± 186	700 ± 14.2
Sinapis alba L.	Yellow mustard seeds	183 ± 2.77	599 ± 29.6	240 ± 4.22
Sinapis alba L.	Black mustard seeds	53.2 ± 1.97	523 ± 15.0	229 ± 14.8
Sophora alopecuroides L.	Fenugreek	39.8 ± 2.33	235 ± 23.8	104 ± 9.41
Syzygium aromaticum (L.) Merr. et L. M. Perry	Male clove (flower)	424 ± 14.9	5453 ± 23.9	3131 ± 177
Syzygium aromaticum (L.) Merr. et L. M. Perry	Female clove (fruit)	485 ± 18.5	6682 ± 68.6	3415 ± 53.1
Thymus vulgaris L.	Thyme	241 ± 14.7	3244 ± 143	743 ± 20.0
Zanthoxylum bungeanum Maxim	Red Chinese prickly ash	168 ± 6.58	1844 ± 29.7	1444 ± 245
Zanthoxylum bungeanum Maxim	Green Chinese prickly ash	193 ± 12.8	1961 ± 35.4	1398 ± 82.6

The data were averages of three measurements with standard deviation. TPC, total phenolic content; FRAP, ferric-reducing antioxidant power; TEAC, trolox equivalent antioxidant capacity.

3.5. TPC of Spice Extracts

The TPC in spice extracts was analyzed through the Folin-Ciocalteu method, and varied from 7.35 to 485 mg GAE/g DW extract powder (Table 4). The highest TPC was observed in the extract of female clove, followed by semen alpiniae katsumadai, male colve, red pepper, and fructus amomi, and the TPC values were 485 ± 18.5, 473 ± 8.67, 424 ± 14.9, 378 ± 3.52, and 360 ± 6.80 mg GAE/g DW extract powder, respectively. Additionally, the lowest TPC was found in the extract of dried lemon (7.35 mg GAE/g DW), consistent with the results of antioxidant activity.

3.6. Correlation Analysis

In order to shed light on the potential antibacterial components in spice extracts, the correlations among antibacterial activity on *S. aureus* (indicated by the DIZ values), antioxidant activity (indicated by the FRAP and TEAC values), and TPC were analyzed. As shown in Table 5, a strong positive correlation was found between TPC and FRAP/TEAC values ($r > 0.9$, $p < 0.01$), suggesting that polyphenols mainly contributed to the antioxidant activity of spice extracts. However, regarding the relationship between DIZ values and FRAP/TEAC values, a significant but weak positive correlation ($r < 0.5$, $p < 0.01$) was observed, indicating that antioxidant activity of spice extracts was only slightly related to their antibacterial capacity. Besides, DIZ values were also found to be weakly correlated with TPC ($r = 0.541$ and 0.568, $p < 0.01$), suggesting that polyphenols were only partly responsible for the antibacterial activity of spice extracts on *S. aureus*, and some other substances should exist in spice extracts to contribute to their overall antibacterial activity.

Table 5. Correlation analysis among antibacterial activity, antioxidant activity, and total phenolic content.

Pearson Correlation Coefficient (r)	DIZ Value (*S. aureus* ATCC 25923)	DIZ Value (*S. aureus* SJTUF 20978)	TPC	FRAP	TEAC
DIZ value (*S. aureus* ATCC 25923)	1	0.956 ($p < 0.001$)	0.541 ($p < 0.001$)	0.466 ($p < 0.001$)	0.448 ($p < 0.001$)
DIZ value (*S. aureus* SJTUF 20978)		1	0.568 ($p < 0.001$)	0.490 ($p < 0.001$)	0.448 ($p < 0.001$)
TPC			1	0.919 ($p < 0.001$)	0.931 ($p < 0.001$)
FRAP				1	0.924 ($p < 0.001$)

Pearson correlation analysis was performed to analyze the relationships among the means of parameters. $p < 0.01$ was defined as statistical significance.

3.7. Principal Component Analysis

To further analyze the relationships among antibacterial activity, antioxidant activity, and TPC of spice extracts and to select suitable spice extracts with good antioxidant and antibacterial activities as food preservative candidates, principal component analysis (PCA) was performed to cluster factors, including TPC, FRAP, TEAC, and DIZ values against both *S. aureus* ATCC 25923 and *S. aureus* SJTUF 20978 (Figure 2). According to the results of Kaiser–Meyer–Olkin (KMO) and Bartlett's test (KMO value = 0.725, $p < 0.05$), as well as the communalities of factors with extraction >0.94, the data met the requirements of PCA. In addition, the cumulative variance contribution rate of the two main components (C1 and C2) extracted was 96.3%, with C1 counting for 55.6% and C2 being 40.7%. According to the rotated component matrix, C1 included the factors TPC, FRAP, and TEAC, suggesting that TPC was closely related to the antioxidant capacity. On the other hand, C2 contained DIZ values of *S. aureus* ATCC 25923 (DIZ1) and drug-resistant *S. aureus* SJTUF 20978 (DIZ2). More interestingly, C1 and C2 were clearly divided into two separate clusters (Figure 2), indicating there was no evident relationship of antibacterial capacity with antioxidant ability and TPC. These results were generally in agreement with those of the correlation analysis.

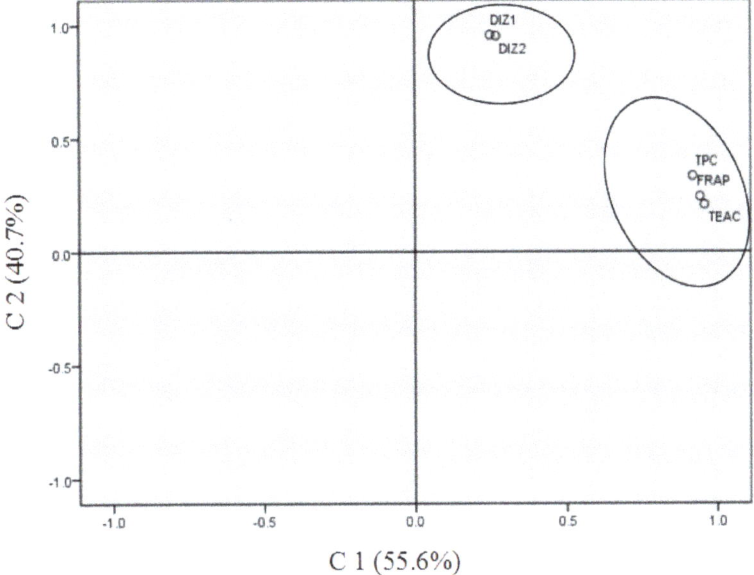

Figure 2. The results of principal component analysis (PCA). C1 included the factors total phenolic content (TPC), ferric-reducing antioxidant power (FRAP), and trolox equivalent antioxidant capacity (TEAC), suggesting that TPC was closely related to the antioxidant capacity. C2 contained DIZ values of *S. aureus* ATCC 25923 (DIZ1) and drug resistant *S. aureus* SJTUF 20978 (DIZ2). C1 and C2 were clearly divided into two separate clusters, suggesting that there was no evident relationship of antibacterial capacity with antioxidant ability and TPC.

Next, the general score (GS) of each sample was calculated based on the factor scores of the two principal components, following the equation GS = (C1 × 0.556 + C2 × 0.407)/0.963, and the results are shown in Table S1. Spice extracts with higher GS usually exhibited higher phenolic contents, antioxidant, and antibacterial activity. Based on the results listed in Table S1, clove extracts, prepared from both fruit and flower of clove, as well as cinnamon showed the highest GS values, indicating that they could be used as potential promising food preservatives by means of reducing microbial contamination and lipid spoilage oxidation simultaneously.

4. Discussion

In this study, 38 out of 67 tested spices displayed various degrees of antibacterial activity against antibiotic-resistant strain of *S. aureus*, while only four were effective against drug-resistant strain of *S. enteritidis*. The antibacterial activity seemed to be bacteria-dependent, and Gram-positive bacteria were more susceptible to the tested spice extracts than Gram-negative bacteria, which was in accordance with many previous studies [31,32]. Different from Gram-positive bacteria, Gram-negative bacteria have an outer membrane rich in lipopolysaccharides, as well as a unique periplasmic space. The complex composition and spatial structure of lipopolysaccharides form a barrier for penetration of antimicrobial agents, besides, the presence of enzymes in periplasmic space may break down intrusive molecules, preventing the antibacterial drugs entering intracellular environment [29]. Additionally, the antibacterial activity of certain spice extracts tested in our study was also reported by previous studies. For instance, chilli, lemongrass, bay leaf, cumin, cinnamon, clove, parsley, basil, sage, thyme, rosemary, and mint, were all demonstrated to show antibacterial capacity against *S. aureus* [33–36]. However, considering the difference in extraction solvent, extraction method, and dosage of samples, it is difficult to directly compare these results with the results of our present study. More importantly, the inhibitory effects of spice extracts on multi-drug resistant bacteria were relatively less reported. Gull et al. revealed that eight drug-resistant bacteria were inhibited by ginger extract at a concentration of 100 mg/mL, with the DIZ ranging from 11 to 15 mm [37]. Mandal et al. reported that the DIZ value obtained from ethanol extracts (20 μL, 10 mg/mL) of cinnamon, clove, and cumin against methicillin-resistant *S. aureus* was in the range of 22–27 mm, 19–23 mm, and 9–15 mm, respectively [38]. Similarly, Revati et al. found that high level gentamicin-resistant enterococci isolates were sensitive to ethanol extracts (50 μL, 100 mg/mL) of cinnamon, ginger, clove, and cumin, with the DIZ values of 31–34, 27–30, 25–26, and 19–20 mm, respectively [14]. Even though, most of the previous investigations were carried out with a limited number of antibiotic-resistant bacterial isolates as well as the tested spice samples, thus the broad antibacterial spectra of spice extracts could not be demonstrated. In addition, we did not set a positive control, such as antibiotics, mainly with two reasons. On the one hand, the antibiotic resistance of bacterial strains used in our study was determined using 11 different antibiotics (Table 2). On the other hand, we were not intended to compare the antibacterial activity of these 67 spice extracts with antibiotics, since the effects of the crude extracts were generally not comparable to pure antibiotics. Besides, in our study, we used the stock concentration of extracts at a relatively high concentration, 100 mg/mL, for the DIZ evaluation, since our samples were dissolved in DMSO, which also possessed an antibacterial effect at a relatively high concentration, such as more than 5%. To rule out the interference of DMSO during subsequent MIC and MBC assays, it was necessary to increase the stock concentration of extracts to reduce the concentration of DMSO in the final working solution of samples.

In our present study, we further chose 11 spices whose DIZ values were higher than 15 mm to verify their antibacterial effects on another ten antibiotic-resistant strains of *S. aureus*, since tested spice extracts exhibited much better antibacterial activity on Gram-positive *S. aureus* than Gram-negative *S. enteritidis*, and found that galangal, fructus galangae, cinnamon, yellow mustard seed, and rosemary overall had the best antibacterial effect, and could probably be developed into antimicrobial agents. Our study may be the first large-scale investigation on the antibacterial effect of spice hydrophilic extracts on antibiotic-resistant bacteria. Therefore, this study can give a clear comparison of the antibacterial activity of spice extracts, especially against antibiotic-resistant bacteria. To provide useful information like safety for further use of these spice extracts, HFF cells were used to evaluate the cytotoxicity of them by MTT assays. All spices except galangal, rosemary, and sage were with low toxicity with LC_{50} values higher than 100 μg/mL. It was worth noting that galangal, which exhibited excellent antibacterial activity among tested spices was also found to show some cytotoxicity against HFF cells in vitro, while its toxicity should also be evaluated in in vivo studies in the future before reaching the conclusion on its toxicity. Discarding crude extracts with good antimicrobial activity only based on the in vitro cytotoxic experiments should involve caution, since cytotoxic compounds might not necessarily be the same antibacterial compounds in some cases [27]; therefore, the main antibacterial and cytotoxic compounds of galangal ethanol extracts should be further isolated and identified in the future before a final conclusion can be made.

In addition to microbial contamination, lipid oxidation is another major cause of food spoilage, therefore, we also measured the antioxidant capacity of 67 spice extracts. The antioxidant activity of tested 67 spice extracts determined by both FRAP and TEAC assays were in the range of 50.3–6682 mmol Fe(II)/g DW and 17.4–3415 mmol trolox/g DW extract powder, respectively. Among them, clove showed the highest antioxidant capacity, even comparable to butylated hydroxyanisole (BHA), an antioxidant commonly applied in food industry preservation due to its excellent hydrogen-donating capacity and metal-chelation ability [39]. Additionally, the results of PCA analysis showed that the extract of clove (both fruit and flower) and cinnamon were spotlighted as potential good candidates as natural food preservatives due to their excellent antibacterial and antioxidant properties. Several other spice extracts like coriander, cinnamon, oregano, mustard, holy basil, and green pepper were also reported to be potent food preservatives [40–44]. Indeed, some studies demonstrated the potential application of clove extracts in raw chicken meat and raw pork during storage to extend shelf-life, in terms of reducing microbes, maintaining natural color, and retarding lipid oxidation [40]. The antimicrobial and antioxidant activities of clove were mainly attributed to the presence of secondary metabolites. A study conducted by Suleiman et al. revealed that the ethanolic extract of clove flower bud appeared to be rich in flavonoids (26.8%), phenolic acid (20.8%), and tannins (4.9%) [45], whose antioxidant effects were already well-known, similar to another phytochemical screening of clove made by Upadhyaya et al. [46]. In addition, the extract of clove flower bud with stronger antimicrobial capacity was also found to exhibit higher phenolic content [47], indicating that phenolic compounds that contributed to the antioxidant activity also displayed antibacterial capacity. Moreover, some components mainly existing in volatile oil also participated in the contribution of antibacterial activity, such as eugenol, isoeugenol, eugenyl acetate, caryophyllene, and humulene. Eugenol was even classified as a substance generally regarded as safe by Food and Drug Administration (FDA). Compared with male clove (flower bud), there were limited studies on female clove (fruit). Although they were derived from the same plant, chemical components were significantly different, and the phytochemicals in clove fruit were identified as eugenol, 2-hydroxy-4, 6-dimethoxy-5-methylacetophenone, and cyclohexane, which might exert antibacterial and antioxidant effects [48].

The antimicrobial activity of spice extracts is mainly attributed to their phytochemicals. Phenolic compounds, such as phenolic acids, flavonoids, and tannins are among the most abundant and widely distributed groups of secondary metabolites in edible plants [49,50]. Moreover, phenolic compounds have been reported to be highly responsible for the antioxidant activity in spices [9], which is also agreement with our results, showing strong correlation between TPC and FRAP/ABTS values ($r = 0.918$ and 0.931, respectively, $p < 0.01$). Thus, TPC can serve as a bridge connecting the antibacterial and antioxidant activity of spice extracts. In a previous study, Shan et al. showed that there was a strong positive linear relationship among antibacterial activity, antioxidant activity, and TPC values in spices [29]. Indeed, in some spices like sage, higher antibacterial activity could be observed in spices containing higher TPC [51]. Moreover, some phenolic compounds identified in spices showed good bacterial inhibitory efficiency. Taking oregano as an example, its antibacterial activity was strongly linked to the presence of phenolic compounds like carvacrol and thymol [52]. Besides, the phenolic compounds identified in many spices like curcumin in turmeric, eugenol in cloves, thymol in thyme, and gingerol in ginger, as well as caffeic acids and ferulic acids in thyme, cinnamon, and galangal have also been demonstrated to exhibit evident antibacterial capacity [8,50,53–56]. Moreover, the number and position of phenolic hydroxyl groups are also considered to be tightly related to the toxicity towards microorganisms [6]. The antibacterial activity of these phenolic compounds involves many modes of action, such as destroying cell membrane morphology, altering membrane fatty acids, depleting proton motive force, causing reactive oxygen damage, impairing enzymatic mechanisms for energy production and metabolism, disrupting normal functionality of proteins, and inhibiting nucleic acid synthesis [6,29,57].

In our study, however, we found a significant but weak correlation of antibacterial activity with TPC and antioxidant activity, indicating that polyphenols were only partially contributed to the antibacterial activity of spice extracts. The Pearson correlation coefficient ($r = 0.541$) tested between TPC and DIZ values of *S. aureus* in our study was overall consistent with a previous study [24], reporting that the TPC of 28 pigmented edible bean coats were weakly correlated ($r = 0.540$) with DIZ values of *S. aureus*. In addition, Weeakkody et al. found a similarly poor correlation ($r^2 < 0.30$) between the antimicrobial activity of seven edible spice extracts and phenolic compound levels [58]. Our study and these studies suggest that in addition to polyphenols, there should be other substances responsible for the overall antibacterial activity of spice extracts. For instance, in our study, although the TPC of galanga was lower than some other spices, its antibacterial activity was highest among tested spices, indicating that other nonphenolic constituents, like 5-hydroxymethyl furfural (accounting for 59.9% in methanol extract), might have the capacity to act as antimicrobial agents [59,60]. Besides, alkaloids, such as piperine from black pepper, were also found to be effective against *Escherichia coli*, *Klebsiella penumonia*, *Salmonella enterica*, and *S. aureus* [61]. Therefore, polyphenols combined with other bioactive compounds should contribute to the overall antibacterial activity of spice extracts.

5. Conclusions

This study investigated systematically the antibacterial activity, antioxidant activity, and TPC of 67 spice extracts. The antibacterial activity of the spice extracts was partially ascribed to polyphenols, while detailed contributions of other antibacterial components should be elucidated in future work. Five selected spice extracts showed the strongest antibacterial activity against different strains of antibiotic-resistant *S. aureus*, and they have potential for use as antibiotic alternatives in animal feeding. Moreover, the clove, exhibiting both excellent antioxidant and antibacterial activities, has great potential as a natural food preservative in the food industry.

Supplementary Materials: The following are available online at http://www.mdpi.com/2076-2607/7/6/157/s1: Table S1: The calculated general score of 67 extracts of spices.

Author Contributions: Conceptualization, R.-Y.G.; Formal analysis, D.Z.; Funding acquisition, R.-Y.G. and H.C.; Investigation, D.Z., A.K.F., G.K. and Q.-Q.Y.; Methodology, D.Z., A.K.F., G.K. and Q.-Q.Y.; Project administration, R.-Y.G. and H.C.; Resources, X.-M.S., C.-L.S., Q.-X.L. and X.-B.X.; Supervision, R.-Y.G.; Writing—original draft, D.Z.; Writing—review and editing, R.-Y.G., X.-M.S., C.-L.S., Q.-X.L., X.-B.X., H.-B.L. and H.C.

Funding: This study was financially supported by the National Key R&D Program of China (2017YFC1600100), the Shanghai Basic and Key Program (18JC1410800), the Shanghai Pujiang Talent Plan (18PJ1404600), the Agri-X Interdisciplinary Fund of Shanghai Jiao Tong University (Agri-X2017004), and the Shanghai Agricultural Science and Technology Key Program (18391900600).

Conflicts of Interest: The authors declare no conflict of interest.

References

1. Liu, Q.; Meng, X.; Li, Y.; Zhao, C.N.; Tang, G.Y.; Li, H.B. Antibacterial and antifungal activities of spices. *Int. J. Mol. Sci* **2017**, *18*, 1283. [CrossRef] [PubMed]
2. Dhara, L.; Tripathi, A. Antimicrobial activity of eugenol and cinnamaldehyde against extended spectrum beta lactamase producing enterobacteriaceae by in vitro and molecular docking analysis. *Eur. J. Integr. Med.* **2013**, *5*, 527–536. [CrossRef]
3. Boskovic, M.; Zdravkovic, N.; Ivanovic, J.; Janjic, J.; Djordjevic, J.; Starcevic, M.; Baltic, M.Z. Antimicrobial activity of Thyme (*Tymus vulgaris*) and Oregano (*Origanum vulgare*) essential oils against some food-borne microorganisms. *Procedia Food Sci.* **2015**, *5*, 18–21. [CrossRef]
4. Farahani, R.K.; Ehsani, P.; Ebrahimi-Rad, M.; Khaledi, A. Molecular detection, virulence genes, biofilm formation, and antibiotic resistance of *Salmonella enterica* serotype enteritidis isolated from poultry and clinical samples. *Jundishapur J. Microbiol.* **2018**, *11*, e69504.
5. Liu, J.X.; Huang, D.F.; Hao, D.L.; Hu, Q.P. Chemical composition, antibacterial activity of the essential oil from roots of *radix aucklandiae* against selected food-borne pathogens. *Adv. Biosci. Biotechnol.* **2014**, *5*, 1043–1047. [CrossRef]
6. Borges, A.; Ferreira, C.; Saavedra, M.J.; Simoes, M. Antibacterial activity and mode of action of ferulic and gallic acids against pathogenic bacteria. *Microb. Drug Resist.* **2013**, *19*, 256–265. [CrossRef]
7. Fowler, Z.L.; Shah, K.; Panepinto, J.C.; Jacobs, A.; Koffas, M.A. Development of non-natural flavanones as antimicrobial agents. *PLoS ONE* **2011**, *6*, e25681. [CrossRef]
8. Nabavi, S.F.; Di Lorenzo, A.; Izadi, M.; Sobarzo-Sanchez, E.; Daglia, M.; Nabavi, S.M. Antibacterial effects of cinnamon: from farm to food, cosmetic and pharmaceutical industries. *Nutrients* **2015**, *7*, 7729–7748. [CrossRef] [PubMed]
9. Gottardi, D.; Bukvicki, D.; Prasad, S.; Tyagi, A.K. Beneficial effects of spices in food preservation and safety. *Front. Microbiol.* **2016**, *7*, 1394. [CrossRef] [PubMed]
10. De Candia, S.; Quintieri, L.; Caputo, L.; Baruzzi, F. Antimicrobial activity of processed spices used in traditional Southern Italian sausage processing. *J. Food Process. Preserv.* **2017**, *41*, e13022. [CrossRef]

11. Irshad, S.; Ashfaq, A.; Muazzam, A.; Yasmeen, A. Antimicrobial and anti-prostate cancer activity of turmeric (*Curcuma longa* L.) and black pepper (*Piper nigrum* L.) used in typical Pakistani cuisine. *Pak. J. Zool.* **2017**, *49*, 1665–1669. [CrossRef]
12. Nassan, M.A.; Mohamed, E.H.; Abdelhafez, S.; Ismail, T.A. Effect of clove and cinnamon extracts on experimental model of acute hematogenous pyelonephritis in albino rats: Immunopathological and antimicrobial study. *Int. J. Immunopathol. Pharmacol.* **2015**, *28*, 60–68. [CrossRef] [PubMed]
13. Naveed, R.; Hussain, I.; Mahmood, M.S.; Akhtar, M. In vitro and in vivo evaluation of antimicrobial activities of essential oils extracted from some indigenous spices. *Pak. Vet. J.* **2013**, *33*, 413–417.
14. Revati, S.; Bipin, C.; Chitra, P.; Minakshi, B. In vitro antibacterial activity of seven Indian spices against high level gentamicin resistant strains of enterococci. *Arch. Med. Sci.* **2015**, *4*, 863–868. [CrossRef]
15. Voukeng, I.K.; Kuete, V.; Dzoyem, J.; Fankam, A.G.; Noumedem, J.A.K.; Kuiate, J.R.; Pages, J.M. Antibacterial and antibiotic-potentiation activities of the methanol extract of some cameroonian spices against Gram-negative multi-drug resistant phenotypes. *BMC Res. Notes* **2012**, *5*, 299. [CrossRef]
16. Siddhartha, E.; Sarojamma, V.; Ramakrishna, V. Bioactive compound rich Indian spices suppresses the growth of β-lactamase produced multidrug ressitant bacteria. *J. Krishna Inst. Med. Sci. Univ.* **2017**, *6*, 10–24.
17. Nikolic, M.; Glamoclija, J.; Ferreira, I.C.F.R.; Calhelha, R.C.; Fernandes, A.; Markovic, T.; Markovic, D.; Giweli, A.; Sokovic, M. Chemical composition, antimicrobial, antioxidant and antitumor activity of *Thymus serpyllum* L., *Thymus algeriensis* Boiss. and Reut and *Thymus vulgaris* L. essential oils. *Ind. Crop. Prod.* **2014**, *52*, 183–190. [CrossRef]
18. Masuda, H.; Hironaka, S.; Matsui, Y.; Hirooka, S.; Hirai, M.; Hirata, Y.; Akao, M.; Kumagai, H. Comparative study of the antioxidative activity of culinary herbs and spices, and hepatoprotective effects of three selected Lamiaceae plants on carbon tetrachloride-induced oxidative stress in rats. *Food Sci. Technol. Res.* **2015**, *21*, 407–418. [CrossRef]
19. Coccimiglio, J.; Alipour, M.; Jiang, Z.H.; Gottardo, C.; Suntres, Z. Antioxidant, antibacterial, and cytotoxic activities of the ethanolic *Origanum vulgare* extract and its major constituents. *Oxid. Med. Cell Longev.* **2016**, 1404505.
20. Mozaffari Nejad, A.S.; Shabani, S.; Bayat, M.; Hosseini, S.E. Antibacterial effect of garlic aqueous extract on *Staphylococcus aureus* in hamburger. *Jundishapur J. Microbiol.* **2014**, *7*, e13134. [CrossRef]
21. Shan, B.; Cai, Y.Z.; Brooks, J.D.; Corke, H. Potential application of spices and herb extracts as natural preservatives in cheese. *J. Med. Food* **2011**, *14*, 284–290. [CrossRef]
22. Shan, B.; Cai, Y.Z.; Brooks, J.D.; Corke, H. Antibacterial and antioxidant effects of five spice and herb extracts as natural preservatives of raw pork. *J. Sci. Food Agric.* **2009**, *89*, 1870–1885. [CrossRef]
23. Chan, C.L.; Gan, R.Y.; Shah, N.P.; Corke, H. Polyphenols from selected dietary spices and medicinal herbs differentially affect common food-borne pathogenic bacteria and lactic acid bacteria. *Food Control* **2018**, *92*, 437–443. [CrossRef]
24. Gan, R.Y.; Deng, Z.Q.; Yan, A.X.; Shah, N.P.; Lui, W.Y.; Chan, C.L.; Corke, H. Pigmented edible bean coats as natural sources of polyphenols with antioxidant and antibacterial effects. *LWT-Food Sci. Technol.* **2016**, *73*, 168–177. [CrossRef]
25. Elshikh, M.; Ahmed, S.; Funston, F.; Dunlop, P.; McGaw, M.; Marchant, R.; Banat, I.M. Resazurin-based 96-well plate microdilution method for the determination of minimum inhibitory concentration of biosurfactants. *Biorechnol. Lett.* **2016**, *38*, 1015–1019. [CrossRef]
26. Mosmann, T. Rapid colorimetric assay for cellular growth and survival: Application to proliferation and cytotoxicity assays. *J. Immunol. Methods* **1983**, *65*, 55–63. [CrossRef]
27. Elisha, I.L.; Botha, F.S.; McGaw, L.J.; Eloff, J.N. The antibacterial activity of extracts of nine plant species with good activity against *Escherichia coli* against five other bacteria and cytotoxicity of extracts. *BMC Complement. Altern. Med.* **2017**, *17*, 133. [CrossRef]
28. Gan, R.Y.; Kuang, L.; Xu, X.R.; Zhang, Y.A.; Xia, E.Q.; Song, F.L.; Li, H.B. Screening of natural antioxidants from traditional Chinese medicinal plants associated with treatment of rheumatic disease. *Molecules* **2010**, *15*, 5988–5997. [CrossRef]
29. Shan, B.; Cai, Y.Z.; Brooks, J.D.; Corke, H. The in vitro antibacterial activity of dietary spice and medicinal herb extracts. *Int. J. Food Microbiol.* **2007**, *117*, 112–119. [CrossRef]

30. Lamola, M.S. Antimicrobial Activity and Cytotoxicity of Extracts and an Isolated Compound from the Edible Plant Grewia Flava against Four Enteric Pathogens. Master's Thesis, University of Pretoria, Pretoria, South Africa, 2015.
31. Benmeziane, F.; Djermoune-Arkoub, L.; Hassan, K.A.; Zeghad, H. Evaluation of antibacterial activity of aqueous extract and essential oil from garlic against some pathogenic bacteria. *Int. Food Res. J.* **2018**, *25*, 561–565.
32. Nagy, M.; Socaci, S.A.; Tofan, M.; Pop, C.; Mureşan, C.; Pop Cuceu, A.V.; Salan, L.; Rotar, A.M. Determination of total phenolics, antioxidant capacity and antimicrobial activity of selected aromatic spices. Bulletin of University of Agricultural Sciences and Veterinary Medicine Cluj-Napoca. *Food Sci. Technol.* **2015**, *72*, 82–85.
33. Awan, U.A.; Ali, S.; Shahnawaz, A.M.; Shafique, I.; Zafar, A.; Khan, M.A.R.; Ghous, T.; Saleem, A.; Andleeb, S. Biological activities of *Allium sativum* and *Zingiber officinale* extracts on clinically important bacterial pathogens, their phytochemical and FT-IR spectroscopic analysis. *Pak. J. Pharm. Sci.* **2017**, *30*, 729–745.
34. Dghaim, R.; Al Sabbah, H.; Al Zarooni, A.H.; Khan, M.A. Antibacterial effects and microbial quality of commonly consumed herbs in Dubai, United Arab Emirates. *Int. Food Res. J.* **2017**, *24*, 2677–2684.
35. Puangpronpitag, D.; Niamsa, N.; Sittiwet, C. Anti-microbial properties of clove (*Eugenia caryophyllum* Bullock and Harrison) aqueous extract against food-borne pathogen bacteria. *Int. J. Pharmacol.* **2009**, *5*, 281–284.
36. Sagdic, O.; Ozkan, G.; Aksoy, A.; Yetim, H. Bioactivities of essential oil and extract of *Thymus argaeus*, Turkish endemic wild thyme. *J. Sci. Food Agric.* **2009**, *89*, 791–795. [CrossRef]
37. Gull, I.; Saeed, M.; Shaukat, H.; Aslam, S.M.; Samra, Z.Q.; Athar, A.M. Inhibitory effect of *Allium sativum* and *Zingiber officinale* extracts on clinically important drug resistant pathogenic bacteria. *Ann. Clin. Microbiol. Antimicrob.* **2011**, *11*, 8. [CrossRef]
38. Mandal, S.; DebMandal, M.; Saha, K.; Pal, N.K. In vitro antibacterial activity of three Indian spices against methicillin-resistant *Staphylococcus aureus*. *Oman Med. J.* **2011**, *26*, 319–323. [CrossRef] [PubMed]
39. Tan, S.H.; Karim, M.R. Effects of three drying treatments on the polyphenol content, antioxidant and antimicrobial properties of *Syzygium aromaticum* Extract. *Chiang Mai J. Sci.* **2018**, *45*, 937–948.
40. Krishnan, K.R.; Babuskin, S.; Babu, P.A.S.; Sasikala, M.; Sabina, K.; Archana, G.; Sivarajan, M.; Sukumar, M. Antimicrobial and antioxidant effects of spice extracts on the shelf life extension of raw chicken meat. *Int. J. Food Microbiol.* **2014**, *171*, 32–40. [CrossRef]
41. Geremew, T.; Kebede, A.; Andualem, B. The role of spices and lactic acid bacteria as antimicrobial agent to extend the shelf life of metata ayib (traditional Ethiopian spiced fermented cottage cheese). *J. Food Sci. Technol.* **2015**, *52*, 5661–5670. [CrossRef]
42. Ghasemzadeh, A.; Jaafar, H.Z.; Rahmat, A. Changes in antioxidant and antibacterial activities as well as phytochemical constituents associated with ginger storage and polyphenol oxidase activity. *BMC Complement. Altern. Med.* **2016**, *16*, 382. [CrossRef] [PubMed]
43. Nugboon, K.; Intarapichet, K. Antioxidant and antibacterial activities of Thai culinary herb and spice extracts, and application in pork meatballs. *Int. Food Res. J.* **2015**, *22*, 1788–1800.
44. Prachayasittikul, V.; Prachayasittikul, S.; Ruchirawat, S.; Prachayasittikul, V. Coriander (*Coriandrum sativum*): A promising functional food toward the well-being. *Food Res. Int.* **2018**, *105*, 305–323. [CrossRef] [PubMed]
45. Suleiman, W.B.; El Bous, M.M.; El Said, M.; El Baz, H. In vitro evaluation of *Syzygium aromaticum* L. ethanol extract as biocontrol agent against postharvest tomato and potato diseases. *Egypt. J. Bot.* **2019**, *59*, 81–94.
46. Upadhyaya, S.; Yadav, D.; Chandra, R.; Arora, N. Evaluation of antibacterial and phytochemical properties of different spice extracts. *Afr. J. Microbiol. Res.* **2018**, *12*, 27–37.
47. El Maati, M.F.A.; Mahgoub, S.A.; Labib, S.M.; Al Gaby, A.M.A.; Ramadan, M.F. Phenolic extracts of clove (*Syzygium aromaticum*) with novel antioxidant and antibacterial activities. *Eur. J. Integr. Med.* **2016**, *8*, 494–504. [CrossRef]
48. Yang, J.Y.; Zhu, C.C.; Gu, L.H.; Hou, H.S. Studies on characteristic fingerprint of *Fructus Caryophylli*. *Tradit. Chin. Drug Res. Clin. Pharmacol.* **2015**, *26*, 226–230.
49. Cueva, C.; Moreno-Arribas, M.V.; Martin-Alvarez, P.J.; Bills, G.; Vicente, M.F.; Basilio, A.; Rivas, C.L.; Requena, T.; Rodriguez, J.M.; Bartolome, B. Antimicrobial activity of phenolic acids against commensal, probiotic and pathogenic bacteria. *Res. Microbiol.* **2010**, *161*, 372–382. [CrossRef] [PubMed]
50. Fankam, A.G.; Kuete, V.; Voukeng, I.K.; Kuiate, J.R.; Pages, J.M. Antibacterial activities of selected Cameroonian spices and their synergistic effects with antibiotics against multidrug-resistant phenotypes. *BMC Complement. Altern. Med.* **2011**, *11*, 104. [CrossRef] [PubMed]

51. Mekinic, I.G.; Skroza, D.; Ljubenkov, I.; Simat, V.; Mozina, S.S.; Katalinic, V. In vitro antioxidant and antibacterial activity of Lamiaceae phenolic extracts: A correlation study. *Food Technol. Biotechnol.* **2014**, *52*, 119–127.
52. Pezzani, R.; Vitalini, S.; Iriti, M. Bioactivities of *Origanum vulgare* L.: An update. *Phytochem. Rev.* **2017**, *16*, 1253–1268. [CrossRef]
53. D'Souza, S.P.; Chavannavar, S.V.; Kanchanashri, B.; Niveditha, S.B. Pharmaceutical perspectives of spices and condiments as alternative antimicrobial remedy. *J. Evid. Based Complement. Altern. Med.* **2017**, *22*, 1002–1010. [CrossRef]
54. Soleimani, V.; Sahebkar, A.; Hosseinzadeh, H. Turmeric (*Curcuma longa*) and its major constituent (curcumin) as nontoxic and safe substances: Review. *Phytother. Res.* **2018**, *32*, 985–995. [CrossRef] [PubMed]
55. Maistro, E.L.; Angeli, J.P.F.; Andrade, S.F.; Mantovani, M.S. In vitro genotoxicity assessment of caffeic, cinnamic and ferulic acids. *Genet. Mol. Res.* **2011**, *10*, 1130–1140. [CrossRef] [PubMed]
56. Martins, N.; Barros, L.; Santos-Buelga, C.; Henriques, M.; Silva, S.; Ferreira, I.C. Decoction, infusion and hydroalcoholic extract of *Origanum vulgare* L.: Different performances regrding bioactivity and phenolic compounds. *Food Chem.* **2014**, *158*, 73–80. [CrossRef] [PubMed]
57. Marchese, A.; Barbieri, R.; Coppo, E.; Orhan, I.E.; Daglia, M.; Nabavi, S.F.; Izadi, M.; Abdollahi, M.; Nabavi, S.M.; Ajami, M. Antimicrobial activity of eugenol and essential oils containing eugenol: A mechanistic viewpoint. *Crit. Rev. Microbiol.* **2017**, *43*, 668–689. [CrossRef]
58. Weerakkody, N.S.; Caffin, N.; Lambert, L.K.; Turner, M.S.; Dykes, G.A. Synergistic antimicrobial activity of galangal (*Alpinia galanga*), rosemary (*Rosmarinus officinalis*) and lemon iron bark (*Eucalyptus staigerana*) extracts. *J. Sci. Food Agric.* **2010**, *91*, 461–468. [CrossRef]
59. Rao, K.; Ch, B.; Narasu, L.M.; Giri, A. Antibacterial activity of *Alpinia galanga* (L) Willd crude extracts. *Appl. Biochem. Biotechnol.* **2010**, *162*, 871–884. [CrossRef] [PubMed]
60. Rini, C.S.; Rohmah, J.; Widyaningrum, L.Y. The antibacterial activity test galanga (*Alpinia galangal*) on the growth of becteria *Bacillus subtilis* and *Escherichia coli*. *IOP Conf. Ser. Mater. Sci. Eng.* **2018**, *420*, 012141. [CrossRef]
61. Zarai, Z.; Boujelbene, E.; Ben Salem, N.; Gargouri, Y.; Sayari, A. Antioxidant and antimicrobial activities of various solvent extracts, piperine and piperic acid from *Piper nigrum*. *LWT-Food Sci. Technol.* **2013**, *50*, 634–641. [CrossRef]

Article

Efficacy of Kombucha Obtained from Green, Oolong, and Black Teas on Inhibition of Pathogenic Bacteria, Antioxidation, and Toxicity on Colorectal Cancer Cell Line

Thida Kaewkod [1,2], Sakunnee Bovonsombut [1,3] and Yingmanee Tragoolpua [1,3,*]

[1] Department of Biology, Faculty of Science, Chiang Mai University, Chiang Mai 50200, Thailand; tda007suju@gmail.com (T.K.); sakunnee.b@cmu.ac.th (S.B.)
[2] The Graduate School, Chiang Mai University, Chiang Mai 50200, Thailand
[3] Center of Excellence in Bioresources for Agriculture, Industry, and Medicine, Department of Biology, Faculty of Science, Chiang Mai University, Chiang Mai 50200, Thailand
[*] Correspondence: yboony150@gmail.com; Tel.: +66-53-941-949-50 (ext. 119); Fax: +66-53-892-259

Received: 16 October 2019; Accepted: 12 December 2019; Published: 14 December 2019

Abstract: Kombucha tea is a refreshing beverage that is produced from the fermentation of tea leaves. In this study, kombucha tea was prepared using 1% green tea, oolong tea, and black tea, and 10% sucrose with acetic acid bacteria and yeast. The pH values of the kombucha tea were found to be in a range of 2.70–2.94 at 15 days of fermentation. The lowest pH value of 2.70 was recorded in the kombucha prepared from black tea. The total acidity of kombucha prepared from black tea was the highest by 16.75 g/L and it was still maintained after heat treatment by boiling and after autoclaved. Six organic acids: glucuronic, gluconic, D-saccharic acid 1,4-lactone, ascorbic, acetic, and succinic acid in kombucha tea were detected by HPLC with the optimization for organic acids detection using isocratic elution buffer with C18 conventional column. The highest level of organic acid was gluconic acid. Kombucha prepared from green tea revealed the highest phenolic content and antioxidation against DPPH radicals by 1.248 and 2.642 mg gallic acid/mL kombucha, respectively. Moreover, pathogenic enteric bacteria: *Escherichia coli*. *E. coli* O157:H7. *Shigella dysenteriae*, *Salmonella* Typhi, and *Vibrio cholera* were inhibited by kombucha and heat-denatured kombucha with diameter of the inhibition zones ranged from 15.0 ± 0.0–25.0 ± 0.0 mm. In addition, kombucha prepared from green tea and black tea demonstrated toxicity on Caco-2 colorectal cancer cells. Therefore, kombucha tea could be considered as a potential source of the antioxidation, inhibition of pathogenic enteric bacteria, and toxicity on colorectal cancer cells.

Keywords: antibacteria; antioxidation; Caco-2 cancer cells; kombucha tea; pathogenic bacteria

1. Introduction

Kombucha tea is a slightly acidic beverage that is produced from the fermentation of tea leaves (*Camellia sinensis*) and infusion with a consortium culture of acetic bacteria including *Acetobacter xylinum*, *A. xylinoides*, or *Bacterium gluconicum* and yeasts such as *Saccharomyces cerevisiae*, *S. ludwigii*, *Zygosaccharomyces bailii*, *Z. rouxii*, *Schizosaccharomyces pombe*, *Torulaspora delbrueckii*, *Brettanomyces bruxellensis*, *B. lambicus*, *B. custersii*, *Candida* sp., or *Pichia membranaefaciens* [1]. Normally, a traditional substrate used in kombucha fermentation is comprised of 10 g/L of black tea infusion that has been sweetened with.5–8% (*w*/*v*) sucrose. Notably, cellulose is produced during the fermentation by *A. xylinum* and appears as a thin film on top of the fermented tea where the cell mass of bacteria and yeast is attached. Yeast and bacteria in kombucha are involved in metabolic activities that utilize substrates in different pathways. Yeast cells hydrolyze sucrose into glucose and hydrolyze fructose using invertase enzymes. Moreover, ethanol is also produced and further utilized by acetic acid bacteria to generate organic acids and other substances such as acetic, gluconic acid, glucuronic acid, citric acid, lactic acid, malic acid, succinic acid, saccharic acid, pyruvic acid, sugars, vitamins, and amino acids [2]. Thus, the pH value of kombucha is known to decrease during the process of fermentation due to the production of organic acids. Kombucha beverages also contain other substances such as phenolic compounds in a quantity of about 30% (*w*/*w*) of the dry mass of the tea leaves [3]. These substances vary considerably depending on the variety of tea and the processing procedures. Moreover, kombucha fermentation is dependent upon the source of kombucha culture (tea fungus), which is affected by a variety of properties for each kombucha beverage. When compared to unfermented tea, the enhanced beneficial activities of kombucha tea indicate that some changes are related to the origin of the microbial community that is present during the fermentation process [4–6].

Non-fermented tea such as green tea contains major polyphenolic catechins, such as epigallocatechin gallate (EGCG), epigallocatechin (EGC), epicatechin gallate (ECG), and epicatechin (EC) [7]. Black tea is considered a fully fermented form of tea since the production process creates a small particle size of the tea leaves and a greater surface area for enzymatic oxidation. During black tea fermentation, quinones react with catechins and produce new compounds; theaflavins and thearubigins [8]. In addition, catechins present in green tea are also partially converted to theaflavins [9]. Oolong tea is classified as semi-fermented tea and prepared during a limited period of oxidation. Thus, the fermentation process of oolong tea is shorter than that of black tea. Notably, oolong tea contains approximately half the amount of catechins when compared to green tea [10]. The compounds that are produced during the production processes of different types of tea markedly affect the composition and total phenolic content that are found in kombucha tea. The major polyphenolic components, catechin and epicatechin, are known to possess antioxidant activity [11]. Moreover, the antimicrobial activity of kombucha from black tea against pathogenic *Vibrio* strains was observed [12].

Although kombucha has been used for long time but scientific report on properties of kombucha has not been clarified. In this study, different biological properties of kombucha tea from various kinds of tea leaves including green, oolong, and black tea were determined for the useful properties of kombucha and application as supplementary beverage for health benefits. Hence, the aim of this study was to investigate the antioxidant and antibacterial properties of kombucha that was obtained from different types of *C. sinensis*, including green tea, oolong tea, and black tea, and the degree of toxicity it displays against the colorectal cancer cell line.

2. Materials and Methods

2.1. Starter Culture of Kombucha Tea

Tea leaves and starter culture were also provided from Tea Gallery Group (Thailand) Co., Ltd., Chiang Mai, Thailand. Kombucha tea was prepared using the same method of the industry standard. The formulation of kombucha tea production consisted of 1% of tea leaf and 10% of sucrose. The starter culture of kombucha tea was used as a consortium culture combining yeast and acetic acid bacteria. The stock culture of kombucha tea contained a total count of bacterial colony at 7.79–7.81 log CFU/mL and the yeast colony was recorded at 7.53–7.75 log CFU/mL.

2.2. Preparation of Kombucha Tea Obtained from Camellia Sinensis Tea Leaves

Kombucha tea was prepared using different types of tea; green, oolong, and black tea. The tea leaves were obtained from the Tea Gallery Group (Thailand) Co., Ltd. Dry tea leaves at 1.0% (*w*/*v*) were added to 500 mL of sterilize distilled water and then boiled for 15 min. Sucrose was filtered into sterilized glass bottles and 10% (*w*/*v*) sucrose was then dissolved in the hot tea. A starter culture made from the fermentation broth of the tea fungus at 10% (*v*/*v*) was inoculated into a mixture of tea and sugar, and it was further incubated at room temperature for 15 days. Subsequently, the kombucha tea was analyzed for its chemical and microbiological properties at 0, 3, 6, 9, 12, and 15 days of fermentation.

2.3. pH and Total Acidity Determination

The pH of kombucha tea was measured using an electronic pH meter (Denver Instruments, Bohemia, NY, USA). The total acidity of kombucha tea was measured according to the procedure of Srihari and Satyanarayana [13]. Kombucha tea was titrated with 0.1 M NaOH. The volume of the NaOH solution was calculated in terms of grams of acetic acid per liter of the sample.

2.4. Total Soluble Solids (°Brix) and Alcohol Content

The total soluble solids (*°Brix*) were measured using a refractometer (RHB-62ATC, JEDTO, Halden, Norway) and the alcohol content of the kombucha tea was measured using an Ebuliometer (Dujardin Salleron, France).

2.5. Microbiological Determination

The total count of yeast cells in the kombucha tea was determined using the spread plate technique on yeast malt (YM) agar with the addition of chloramphenicol. Determination of the total count of acetic acid bacteria was performed on yeast peptone mannitol (YPM) agar with amphotericin B (CAISSON, USA). The plates were incubated at 30 °C, for at least 3 days [14].

2.6. HPLC Analysis of Organic Acids

The predominant organic acids found in kombucha tea were determined by high performance liquid chromatography (HPLC). Six organic acids including glucuronic acid (Sigma-Aldich, Darmstadt, Germany), gluconic acid (Merck, Darmstadt, Germany), D-saccharic acid 1,4-lactone (Sigma-Aldich, Germany), acetic acid (Merck, Germany), ascorbic acid (Merck, Germany), and succinic acid (Merck, Germany) in kombucha tea were optimized for detection by isocratic HPLC systems with a conventional C18 column. The kombucha tea samples were filtered through a 0.22 µm sterile microfilter and 50 µL of the filtrate was injected into the HPLC system (Agilent technologies 1200 series, Santa Clara, CA, USA). The C-18 column (4.6 × 150 mm, 5 µm; GL Sciences, Tokyo, Japan) employing a UV detector (210 nm) was used for the analysis. Moreover, the HPLC system was controlled with a flow rate of 0.8 mL per minute and a running time of 40 min at 25 °C. Six organic acids in kombucha tea were separated by 20 mM KH_2PO_4 elution buffer at pH 2.4 and the standards of organic acids were used for comparison with kombucha tea. The peak area of each organic acid was calculated by Agilent ChemStation level-5 program, and then standard graphs of each organic acid were generated. Thus, the content of each organic acid in kombucha tea was calculated from the standard graph of each organic acid.

2.7. DPPH Radical Scavenging Assay of Kombucha Tea

The radical scavenging activity of kombucha tea was determined against DPPH free radical content [15]. The kombucha tea was diluted with methanol by 10-fold serial dilution. Each concentration of the kombucha tea (0.5 mL) was incubated with 1.5 mL of 0.1 mM DPPH solution (Sigma-Aldich, Germany). Absorbance at 517 nm was measured by spectrophotometer (Genesys 20, Thermo Scientific, Dreieich, Germany) after the solution mixture was kept in the dark at room temperature for 20 min. Methanol was used as a blank solution, while DPPH without kombucha tea was used as a control. The absorbance of the DPPH solution A1, and the absorbance of kombucha tea mixed with the DPPH solution A2, were measured. The percentage of DPPH free radical inhibition was calculated as follows: percentage inhibition = {(A1–A2)/A1} × 100.

The antioxidant activity of kombucha tea was assessed by comparing the samples to standard gallic acid and was expressed as milligrams of gallic acid per milliliter of kombucha tea (mg GAE/mL kombucha tea).

2.8. Total Phenolic Content of Kombucha Tea

Total phenolic content of kombucha tea was determined by Folin-Ciocalteu reagent [16]. Kombucha tea (250 µL) was mixed with 125 µL of 50% Folin-Ciocalteu reagent (Merck, Germany) and 250 µL of 95% ethanol. The mixture was incubated in the dark at room temperature for 5 min. Subsequently, 250 µL of 5% sodium carbonate was then added, and the mixture was incubated in the dark at room temperature for 60 min. Blue molybdenum–tungsten complex was formed and detected at 725 nm by spectrophotometer (Genesys 20, Thermo Scientific, Germany). Total phenolic content was calculated by comparing the substance to standard gallic acid and was expressed as milligrams of gallic acid per milliliter of kombucha tea (mg GAE/mL kombucha tea).

2.9. Pathogenic Enteric Bacteria Used in the Study

The pathogenic enteric bacteria that were used for the antimicrobial activity test included *E. coli* O157:H7 DMST 12743, *Shigella dysenteriae* DMST 1511 and *Salmonella* Typhi DMST 22842. *Escherichia coli* and *Vibrio cholera* were kindly obtained from the Microbiology Section, Department of Medical Technology, Faculty of Associated Medical Science, Chiang Mai University, Chiang Mai, Thailand. The bacterial strains were stored in glycerol stock at −20 °C and then grown on Mueller–Hinton (MH) agar (Difco™, Detroit, MI, USA) plates at 37 °C for 18–24 h.

2.10. Antimicrobial Activity of Kombucha Tea

A single colony of the tested bacteria; *Escherichia coli*, *E. coli* O157:H7, *Shigella dysenteriae*, *Salmonella* Typhi, and *Vibrio cholera*, was transferred into Mueller–Hinton (MH) broth (Difco™, USA) and incubated at 37 °C for 18–24 h. The antimicrobial activity of kombucha tea after 15 days of fermentation was investigated using the agar well diffusion method. The turbidity of the bacterial culture was adjusted to McFarland standard no. 0.5 and swabbed on Mueller–Hinton (MH) agar (Difco™, USA). Wells of 10 mm in diameter were prepared on the agar plate with a sterile cork borer. Kombucha samples were sterilized by filtering them through a sterile microfilter (0.22 µm pore size), and they were then transferred into the wells in the agar plates. The plates were further incubated at 37 °C for 18–24 h. The zones of bacterial growth inhibition were then determined [17,18].

The antimicrobial activity of kombucha tea was compared to the non-fermented tea, and acetic acid was recorded at the same concentration of the kombucha tea after 15 days of fermentation. Moreover, kombucha tea was neutralized at pH 7.0 by adjusting the pH with 1 M HCl or 1 M NaOH. Heat denatured kombucha tea was prepared by treatment at 100 °C for 20 min and at 121.5 °C, at 15 pounds per square inch for 15 min by being autoclaved. After treatment, the kombucha tea was then sterilized by filtration and tested for its antimicrobial activity, as has been described previously.

2.11. Cytotoxicity Test of Kombucha Tea on Human Colorectal Carcinoma (Caco-2) and NIH/3T3 Cells

The cytotoxicities of kombucha tea were tested using MTT assay. NIH/3T3 cells were used as normal cell control. Human colorectal carcinoma (Caco-2) and NIH/3T3 cells were cultured in Dulbecco's modified Eagle medium (DMEM; Gibco, Grand Island, NY, USA) that had been supplemented with 10% (*v*/*v*) heat-inactivated fetal bovine serum (HyClone™, Pittsburgh, PA, USA), 100 Units/mL penicillin and 100 µg/mL streptomycin (CAISSON, Smithfield, UT, USA). After incubation at 37 °C in a 5% CO_2 incubator (SHEL LAB, USA), the cells were washed twice with phosphate buffer saline (PBS, pH 7.4) and trypsinized with 0.05% (*v*/*v*) trypsin-EDTA solution (CAISSON, USA). The Caco-2 and NIH/3T3 cells were plated in 96-well plates and incubated at 37 °C in a 5% CO_2 incubator for 24 h. After incubation, each concentration of kombucha tea was then added. The plates were incubated at 37 °C in a 5% CO_2 incubator for 48 h. The MTT solution (Bio Basic Inc., Amherst, NY, USA) was then added and the solution was incubated for 4 h. Finally, blue formazan crystals were dissolved with dimethyl sulfoxide and the absorbance was measured at 540 and 630 nm by micro plate reader (EZ

Read 2000, Biochrom, Cambridge, UK). The percentage of cell viability was calculated by comparing the relevant values to the cell control [19].

2.12. Statistical Analysis

All experiments were performed in three independent treatments. All data acquired from the treatments and the control groups were compared and analyzed and are presented as mean ± SD using *t*-test and ANOVA analysis.

3. Results

3.1. Appearance of Kombucha Tea during Fermentation

Kombucha tea was prepared from green, oolong, and black teas using 1% (*w*/*v*) tea leaves and 10% (*v*/*v*) fermentation broth in the preparation of tea fungus as a starter culture. After 15 days of fermentation, the kombucha tea prepared from black tea appeared as a dark-brown color, while the kombucha tea that was prepared from the green tea and oolong tea displayed a brown color. During the fermentation period of kombucha preparation over 3–15 days, a thin film appeared on top of the culture medium, which was identified as the cellulose produced by the acetic acid bacteria (Figure 1).

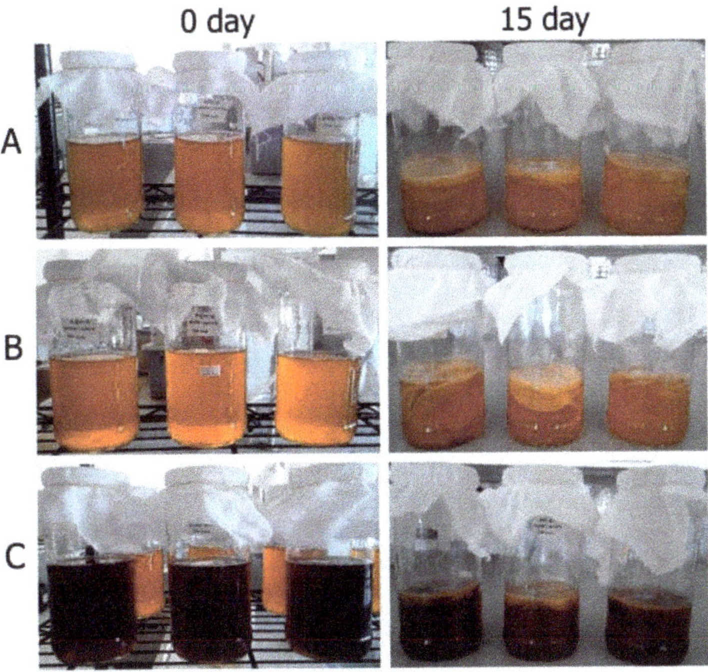

Figure 1. Appearance of kombucha prepared from green tea (**A**), oolong tea (**B**), and black tea (**C**) at the beginning of the fermentation process and after 15 days of fermentation.

3.2. Microbial and Chemical Changes that Occur During Kombucha Tea Fermentation

Microorganisms in kombucha tea utilize a carbon source and begin to produce cellulose which appears as a thin layer on top of the culture medium. The different types of tea; green, oolong, and black tea, that were used in this study did not affect the growth of the tea fungus. The total counts of bacterial and yeast cells during kombucha tea fermentation are shown in Figure 2. After inoculation of the 10% (*v*/*v*) starter culture, total counts of acetic acid bacterial colonies at the beginning of the kombucha tea fermentation process for green, oolong, and black teas were 5.72, 5.81, and 5.54 log CFU/mL, respectively. Moreover, the total counts of yeast cells at the beginning of kombucha tea fermentation were 5.68, 5.89, and 5.79 log CFU/mL in the kombucha tea prepared from green, oolong, and black teas, respectively. After increasing the fermentation time, the number of bacterial and yeast cells were found to be significantly higher than when observed at the beginning of the fermentation period (Figure 2A,B). After 15 days of fermentation, a total count of 7.80 log CFU/mL was found in the kombucha that had been prepared from green tea and black tea. The acetic acid bacterial colonies were significantly higher in the kombucha prepared from green and black teas than in the kombucha prepared from oolong tea (7.54 log CFU/mL). In addition, the total counts of the yeast cells in the kombucha tea prepared from green tea (7.64 log CFU/mL), oolong tea (7.74 log CFU/mL) and black tea (7.78 log CFU/mL) were not significantly different in each type of kombucha tea.

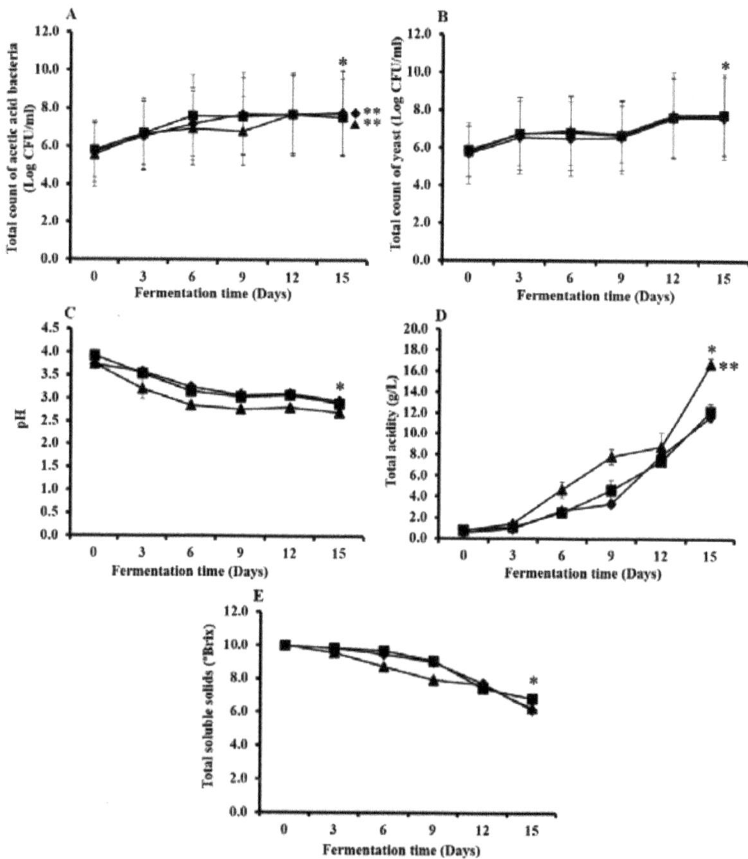

Figure 2. Alteration of the total count of acetic acid bacteria (**A**), total count of yeast cells (**B**), pH (**C**), total acidity (**D**), and total soluble solids (**E**) during fermentation of kombucha prepared from green tea (◆), oolong tea (■), and black tea (▲). * The values were significantly different between the beginning of the fermentation and the end of 15 days of the fermentation period ($p < 0.05$). ** The values were significantly different for each type of kombucha tea at the end of 15 days of fermentation ($p < 0.05$). The results are presented as mean ± SD of three independent experiments.

The blank control without acetic acid and yeast was performed. However, after incubation for 2–3 days the contamination from other microorganisms was presented because the blank control showed pH around 4.57–5.25. However, pH of 3.73–3.92 was determined after inoculation of starter culture at 0 day of fermentation time (Figure 2C). This acidic condition of kombucha tea inhibited other contaminated microorganisms in kombucha tea.

The alteration of pH values during kombucha tea fermentation with significantly different initial pH values is shown in Figure 2C. At the end of the 15-day fermentation period, the pH value of kombucha tea that had been prepared from black tea was the lowest at a pH value of 2.70. Kombucha that was prepared from green tea and oolong tea revealed pH values of 2.94 and 2.89, respectively. On the other hand, changes in titratable acidity that occurred during the fermentation process were significantly increased, which indicated a concentration of these organic acids (Figure 2D). The total acidity of the kombucha prepared from black tea was significantly higher (16.75 g/L) than that of the kombucha prepared from oolong (12.24 g/L) and green teas (11.72 g/L). In contrast, total soluble solids of kombucha prepared from green, oolong and black tea were significantly decreased from 10 °*Brix* to 6 °*Brix* at 15 days of fermentation (Figure 2E). In addition, no alcohol content was detected during the process of kombucha tea fermentation.

3.3. Organic Acids in Kombucha Tea

The organic acids in kombucha tea were analyzed by HPLC assay. The HPLC system was optimized for the detection of several organic acids in kombucha tea with a conventional C18 column. HPLC conditions were optimized and 20 mM KH_2PO_4 with a pH value of 2.4 in the isocratic elution buffer was used with a 210 nm UV detector. Six organic acids including glucuronic acid, gluconic acid, D-saccharic acid 1,4-lactone (DSL), acetic acid, ascorbic acid, and succinic acid were clearly separated. Chromatograms of six standard organic acids, including glucuronic acid, gluconic acid, DSL, acetic acid, ascorbic acid, and succinic acid, were eluted at different retention times (Figure 3).

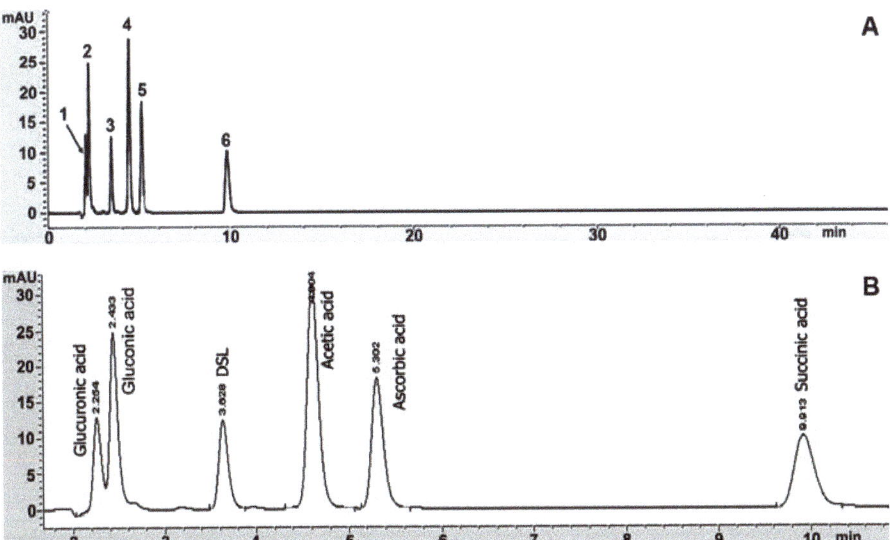

Figure 3. HPLC chromatograms of running time 40 min (**A**) and the six organic acids (**B**) are presented as standards; glucuronic acid (Peak 1), gluconic acid (Peak 2), D-saccharic acid 1,4-lactone (DSL) (Peak 3), acetic acid (Peak 4), ascorbic acid (Peak 5), and succinic acid (Peak 6).

After 15 days of kombucha fermentation, the amounts of each type of organic acid that were present in the kombucha prepared from green, oolong, and black teas were analyzed by HPLC. The chromatograms and production values of organic acids at 15 days of fermentation are shown in Figure 4 and Table 1. The highest content of gluconic acid was found in the kombucha that was prepared from green, oolong, and black teas, followed by acetic acid, DSL, ascorbic acid, and glucuronic acid. In contrast, succinic acid was detected in the kombucha that had been prepared from black tea, but it was not detected in the kombucha prepared from green tea and oolong tea. Therefore, kombucha prepared from black tea revealed the highest amounts of organic acid content: glucuronic acid (1.58 g/L), gluconic acid (70.11 g/L), DSL (5.23 g/L), ascorbic acid (0.70 g/L), acetic acid (11.15 g/L), and succinic acid (3.05 g/L), when compared to both oolong tea and green tea.

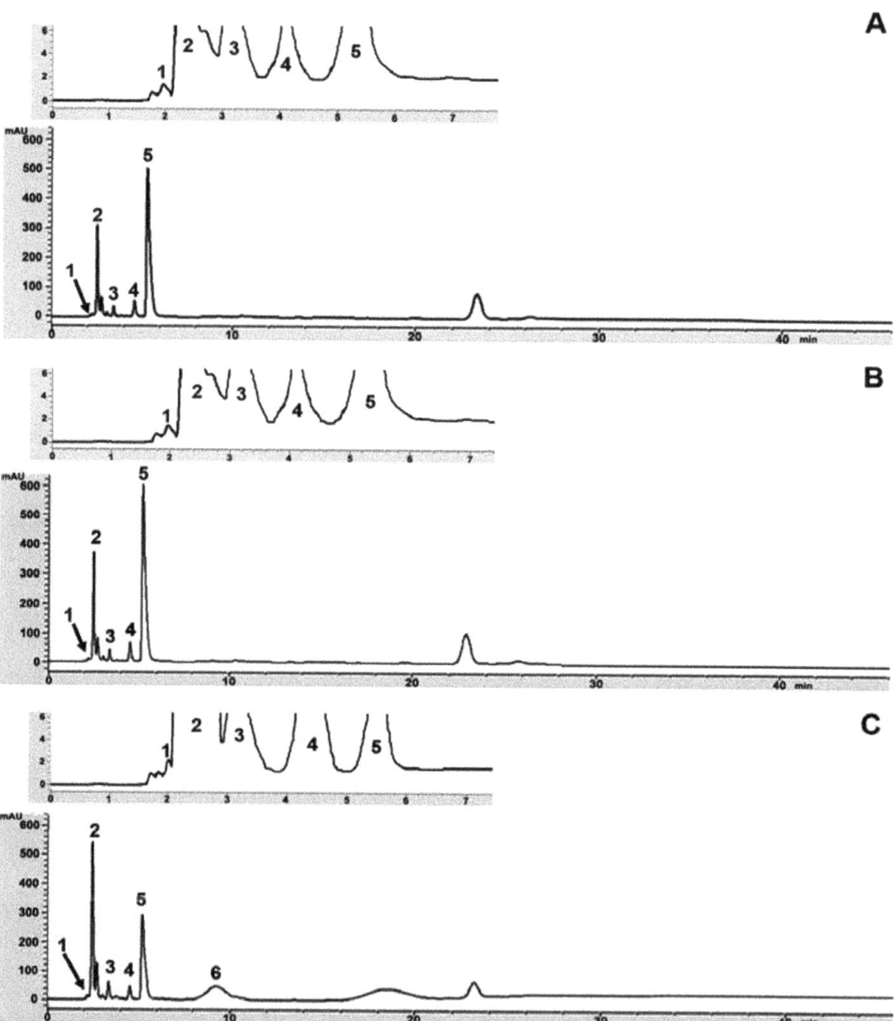

Figure 4. HPLC chromatograms of kombucha prepared from green tea (**A**), oolong tea (**B**), and black tea (**C**) at 15 days of fermentation. Peak (**1**)—glucuronic acid; Peak (**2**)—gluconic acid; Peak (**3**)—D-saccharic acid 1,4-lactone (DSL); Peak (**4**)—acetic acid; Peak (**5**)—ascorbic acid; Peak (**6**)—succinic acid, respectively.

Table 1. Organic acid content of kombucha tea prepared from green tea, oolong tea, and black tea at 15 days of the fermentation process.

Kombucha	Organic Acids Content (g/L)					
	Glucuronic	Gluconic	DSL	Ascorbic	Acetic	Succinic
Green tea	1.37 ± 0.01 [c]	41.42 ± 0.02 [c]	3.44 ± 0.03 [c]	0.61 ± 0.00 [a]	10.42 ± 0.00 [b]	ND [d]
Oolong tea	0.07 ± 0.01 [b]	48.75 ± 0.03 [b]	4.02 ± 0.02 [b]	0.60 ± 0.00 [a]	10.48 ± 0.00 [ab]	ND [d]
Black tea	1.58 ± 0.01 [a]	70.11 ± 0.01 [a]	5.23 ± 0.01 [a]	0.70 ± 0.00 [a]	11.15 ± 0.00 [a]	3.05 ± 0.01 [a]

[a, b, c] The data of different superscript letters (a, b, c) are presented as mean ± SD of three independent experiments and revealed significantly different values for each type of kombucha tea ($p < 0.05$). [d] ND: organic acid in kombucha tea was not detected.

3.4. DPPH Scavenging Ability and Total Phenolic Content during Kombucha Fermentation

The DPPH scavenging ability of kombucha prepared from green, oolong, and black tea significantly increased during the fermentation process to 2.642, 2.582, and 0.435 mg GAE/ml kombucha tea, respectively (Figure 5). The maximum increase of DPPH scavenging activity was presented after 15 days in the kombucha that was prepared from green tea. Kombucha prepared from oolong tea showed the highest DPPH scavenging activity after 9 days of fermentation and was remarkably stable after 12–15 days of fermentation. Moreover, the DPPH scavenging ability of the kombucha prepared from black tea was at the highest incremental value after 3 days of the fermentation process.

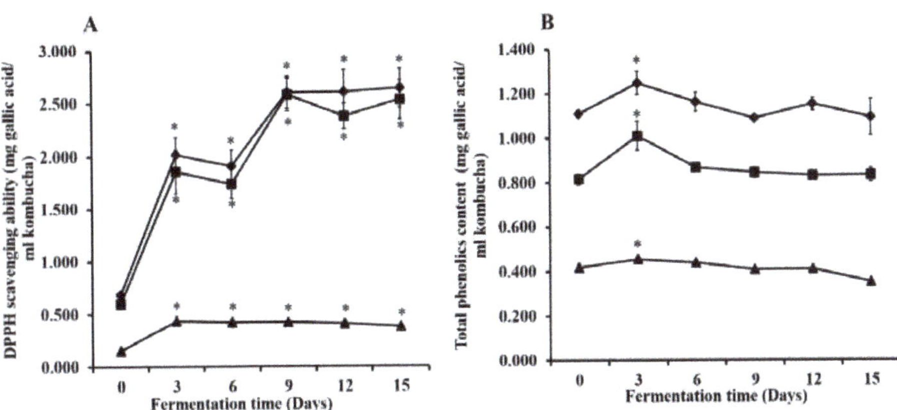

Figure 5. DPPH scavenging ability (**A**) and total phenolic content (**B**) during fermentation of kombucha prepared from green tea (◆), oolong tea (■), and black tea (▲). * The values were significantly different between the beginning of the fermentation and on each day of the fermentation period ($p < 0.05$). The results are presented as mean ± SD of three independent experiments.

Total phenolic content is shown in Figure 5. After fermentation for three days, the kombucha that was prepared from green, oolong, and black teas revealed maximum levels of total phenolic contents at 1.248, 1.011, and 0.455 mg GAE/mL kombucha tea, respectively. After 3–6 days of fermentation, the amounts of phenolic compounds decreased and then appeared to be stable at 15 days of the fermentation process. Thus, kombucha prepared from green tea revealed the highest amounts of antioxidant activity and total phenolic content.

3.5. Total Acidity and Antibacterial Activity of Kombucha Tea

The antibacterial activities of the kombucha tea that was fermented under different conditions were investigated against pathogenic enteric bacteria. The fermented tea was tested at 15 days in terms of acidity and then after neutralization (pH 7.0). Moreover, the samples were heated to analyze the thermostability of the active components. Kombucha tea prepared from green, oolong, and black tea showed antibacterial activity against all tested enteric bacteria; *Escherichia coli*, *E. coli* O157: H7 DMST 12743, *Shigella dysenteriae* DMST 1511, *Salmonella* Typhi DMST 22842, and *Vibrio cholerae*. The diameter of the inhibition zones of the kombucha prepared from green tea ranged from 20.0 ± 0.0–24.7 ± 0.6 mm. The diameter of the inhibition zones of kombucha prepared from oolong tea ranged from 19.3 ± 0.6–24.7 ± 0.6 mm and the diameter of the inhibition zones of black tea ranged from 20.0 ± 0.0–21.3 ± 0.6 mm. The antibacterial activity of each type of kombucha tea was similar to that of the acetic acid at the same total acidity content of each type of kombucha tea. In this study, gentamicin (1 mg/mL) was used as a positive control and presented the diameters of the inhibition zones ranged from 19.3 ± 0.6–23.3 ± 0.6 mm. after testing with *E. coli*, *E. coli* O157:H7, *S. dysenteriae*, *S.* Typhi, and *V. cholera* (Table 2).

Table 2. Antibacterial activity of kombucha prepared from green tea, oolong tea, and black tea against pathogenic enteric bacteria.

Type of *Camellia Sinensis*	Tested Extract	[a] Inhibition Zone Diameter (mm) of Target Bacteria				
		E. coli	*E. coli* O157: H7	*S. Dysenteriae*	*S.* Typhi	*V. Cholera*
Green tea	[b] Fermented tea	24.7 ± 0.6	24.3 ± 0.6	21.7 ± 0.6	23.7 ± 0.6	20.0 ± 0.0
	[c] Unfermented tea	0	0	0	0	0
	[d] Neutralized kombucha	0	0	0	0	0
	[e] Heat-denatured Kombucha M1	25.0 ± 0.0	23.0 ± 0.0	21.0 ± 0.0	23.0 ± 0.0	19.0 ± 0.0
	[f] Heat-denatured Kombucha M2	15.3 ± 0.6	22.3 ± 0.6	19.7 ± 0.6	20.7 ± 0.6	17.3 ± 0.6
	[g] Acetic acid (11.72 g/L)	20.3 ± 0.6	22.0 ± 0.0	20.0 ± 0.0	21.0 ± 0.0	22.0 ± 0.0
Oolong tea	Fermented tea	23.7 ± 0.6	20.3 ± 0.6	19.3 ± 0.6	24.7 ± 0.6	20.0 ± 0
	Unfermented tea	0	0	0	0	0
	Neutralized kombucha	0	0	0	0	0
	Heat-denatured Kombucha M1	20.0 ± 0.0	19.3 ± 0.6	19.3 ± 0.6	24.0 ± 0.0	16.7 ± 0.6
	Heat-denatured Kombucha M2	15.7 ± 0.6	15.0 ± 0.0	17.0 ± 0	19.7 ± 0.6	15.7 ± 0.6
	Acetic acid (12.24 g/L)	21.0 ± 0.0	20.0 ± 0.0	21.0 ± 1.0	20.3 ± 0.6	22.0 ± 0.0
Black tea	Fermented tea	21.0 ± 0.0	21.3 ± 0.6	21.0 ± 0.0	20.0 ± 0.0	21.0 ± 0.0
	Unfermented tea	0	0	0	0	0
	Neutralized kombucha	0	0	0	0	0
	Heat-denatured Kombucha M1	20.3 ± 0.6	20.3 ± 0.6	20.0 ± 0.0	18.0 ± 0.0	20.0 ± 0.0
	Heat-denatured Kombucha M2	18.0 ± 0.0	16.3 ± 0.6	19.7 ± 0.6	17.3 ± 0.6	18.0 ± 0.0
	Acetic acid (16.75 g/L)	20.0 ± 0.0	20.0 ± 0.0	20.7 ± 0.6	21.0 ± 0.0	21.7 ± 0.6
Gentamycin (1 mg/mL)		20.3 ± 0.6	19.3 ± 0.6	23.3 ± 0.6	22.3 ± 0.6	20.0 ± 0.0

[a] Inhibition zone diameter is presented as mean ± SD of three independent experiments. [b] Fermented tea: kombucha tea fermented at 15 days without any treatment. [c] Unfermented tea: broth culture contained only 1% of tea leaves. [d] Neutralized kombucha: kombucha tea neutralized with NaOH (1 M) at pH 7.0. [e] Heat-denatured kombucha M1: kombucha tea treated at 100 °C for 20 min. [f] Heat-denatured kombucha M2: kombucha tea treated at 121.5 °C for 15 min. [g] Acetic acid: acid prepared at the same total acidity content of kombucha for each type of tea leaf.

On the other hand, the antibacterial activity of kombucha after neutralization to pH 7.0 did not reveal any inhibition of all of the tested enteric bacteria. Kombucha tea was also treated by being boiled (100 °C for 20 min) and autoclaved (121.5 °C for 15 min) to determine the activity of the thermostability of its components. The total acidity of the kombucha tea prepared from green, oolong, and black teas were titrated after thermal treatments, while the total acidity of kombucha tea prepared from green, oolong, and black teas after treatment by boiling was 10.93, 11.61, and 15.67 g/L, respectively. In addition, the total acidity of kombucha tea from green, oolong and black teas after treatment by being autoclaved was 8.12, 9.16, and 12.16 g/L, respectively. Thus, the total acidity level of kombucha tea after treatment by boiling was maintained since the total acidity was significantly higher than that of the kombucha tea that was treated by being autoclaved. The results indicated that the antibacterial activity of kombucha tea after treatment by boiling revealed significantly higher inhibitory activity on the tested enteric bacteria than in the kombucha tea that had been treated by being autoclaved (Table 2).

3.6. *Cytotoxicity Test of Kombucha Tea Against Caco-2 and NIH/3T3 Cells*

The inhibition of the kombucha in different types of tea leaves was investigated against Caco-2 cancer cells and NIH/3T3 cells by MTT assay (Table 3). Kombucha tea prepared from green, oolong and black teas at 15 days of fermentation inhibited Caco-2 cancer cells with 50% inhibitory concentrations (IC_{50}) of 2.603%, 1.899%, and 6.077%, respectively. Moreover, Kombucha from green, oolong and black teas inhibited NIH/3T3 cells with 50% inhibitory concentrations (IC_{50}) of 2.851%, 1.922%, and 6.697%, respectively.

Table 3. Cytotoxicity of kombucha prepared from green, oolong, and black teas on Caco-2 colorectal cancer and NIH/3T3 cells.

Type of *Camellia Sinensis*	Tested Extract	[a] IC_{50} (%)	
		Caco-2 Cells	NIH/3T3 Cells
Green tea	[b] Fermented tea	2.603 ± 0.072 *	2.851 ± 0.052 *
	[c] Unfermented tea	3.661 ± 2.228	3.819 ± 0.122
	[d] Neutralized kombucha	8.621 ± 0.685	8.915 ± 0.715
	[e] Heat-denatured kombucha M1	1.758 ± 0.065 *	1.916 ± 0.011*
	[f] Heat-denatured kombucha M2	1.309 ± 0.117	1.518 ± 1.202
	[g] Acetic acid (11.72 g/L)	33.155 ± 2.834	35.206 ± 0.625
Oolong tea	Fermented tea	1.899 ± 0.242	1.922 ± 0.186
	Unfermented tea	2.097 ± 0.305	2.519 ± 0.024
	Neutralized kombucha	4.194 ± 0.081	4.671 ± 1.205
	Heat-denatured kombucha M1	1.179 ± 0.041	1.219 ± 0.012
	Heat-denatured kombucha M2	1.515 ± 0.138	1.641 ± 0.220
	Acetic acid (12.24 g/L)	12.610 ± 0.341	12.817 ± 0.452
Black tea	Fermented tea	6.077 ± 0.222 *	6.697 ± 0.030 *
	Unfermented tea	6.082 ± 0.191	6.244 ± 0.659
	Neutralized kombucha	16.339 ± 0.833	17.066 ± 1.568
	Heat-denatured kombucha M1	6.083 ± 0.053 *	6.702 ± 0.156 *
	Heat-denatured kombucha M2	5.699 ± 0.008	5.819 ± 0.209
	Acetic acid (16.75 g/L)	8.720 ± 0.047	8.810 ± 0.142

[a] Results are presented as mean ± SD of three independent experiments. [b] Fermented tea: kombucha tea fermented at 15 days without any treatment. [c] Unfermented tea: broth culture contained only 1% of tea leaves. [d] Neutralized kombucha: kombucha tea neutralized with NaOH (1 M) at pH 7.0. [e] Heat-denatured kombucha M1: kombucha tea treated at 100 °C for 20 min. [f] Heat-denatured kombucha M2: kombucha tea treated at 121.5 °C for 15 min. [g] Acetic acid: acid prepared at the same total acidity content of kombucha for each type of tea leaf. * The values were significantly different between the cytotoxicity of Caco-2 cells and NIH/3T3 cells ($p < 0.05$).

The effect of kombucha tea on cancer cells showed cytotoxicity values higher than cytotoxicity on normal cells. However, kombucha tea prepared from green tea and black tea significantly showed toxicity against cancer cells when compared to normal cells. Moreover, cytotoxicity of kombucha prepared from green tea and black tea after boiling (100 °C for 20 min) also showed significantly toxicity against cancer cells.

Moreover, all unfermented tea containing 1% (*w*/*v*) of tea leaf-infusion also displayed toxicity toward Caco-2 cancer cells. In contrast, kombucha that was neutralized with 1 M NaOH at a pH value of 7.0 revealed levels of toxicity on Caco-2 cells that were lower than those in the heat-denatured kombucha. Acetic acid was used for acidic activity control and was prepared with the same total acidity content of kombucha for each type of tea. After treatment of Caco-2 cancer cells with acetic acid, IC_{50} values of the kombucha prepared from green, oolong, and black teas were 33.155%, 12.610%, and 8.720%, respectively. Thus, the toxicity of the acetic acid on cancer cells was lower than in the fermented kombucha that had been prepared from each type of tea leaves.

4. Discussion

Kombucha tea was prepared from green, oolong, and black teas. The microbial community and chemical properties presented during kombucha fermentation were demonstrated in this study. In this study, the starter culture of kombucha tea was used as a symbiosis culture of acetic acid bacteria including *Acetobacter xylinum* and yeast cells. Therefore, cellulose production from *A. xylinum* was present on the surface of the kombucha tea during 3–15 days of fermentation. The production of the cellulose of this species is important for kombucha tea fermentation since a floating cellulose pellicle provides benefits to microorganisms by enhancement of the association between the bacteria and yeast cells and allows for the exposure of the microorganisms to atmospheric oxygen [20]. Moreover, caffeine and other related xanthines that are found in tea showed the ability to stimulate cellulose production by way of the bacteria that is present [3].

Normally, kombucha tea was completely fermented when fermentation time was 12–21 days and the inoculation size of starter culture was 5–10%. Thus, a period of kombucha fermentation of 15 days was selected in this study since the highest level of acidity was demonstrated during this fermentation process [11,21–23]. Moreover, yeast cells are known to convert sucrose into glucose and fructose by invertase enzymes. Additionally, ethanol was also produced. Glucose in kombucha was further utilized by acetic acid bacteria to produce gluconic acid, whereas ethanol was used to produce acetic acid. The different types of tea: green, oolong, and black tea that were used in this study did not affect the growth of the tea fungus. After increasing the fermentation time, the total counts of acetic acid bacterial and yeast cells were found to be significantly higher than were observed at the beginning of the fermentation period. After 15 days of fermentation, the total counts of acetic acid bacteria in the kombucha prepared from green tea and black tea were significantly higher than in the kombucha prepared from oolong tea. Additionally, the total counts of yeast cells in each kombucha tea were not significantly different. Therefore, the increasing amounts of bacteria present in kombucha during the fermentation process is dependent upon the different types of tea that are used as a substrate [3].

The pH values of the kombucha prepared from green, oolong, and black teas decreased due to the production of organic acids during the fermentation process. The lowest pH value was recorded in the kombucha that had been prepared from black tea, followed by that of oolong tea and green tea. The total acidity of kombucha prepared from black tea showed the highest total amount of acids, which related to the low pH value of the kombucha. Moreover, microorganisms were able to use sucrose as the major substrate to produce many organic acids that resulted in the decrease of total soluble solids during fermentation of each kombucha tea. These different values of pH and total acidity correspond to the different types of tea used and affected the degree of organic acid production. Notably, green tea is considered a non-fermented type of tea, whereas black tea and oolong tea are classified as fully fermented tea and semi-fermented tea, respectively. The process of tea preparation resulted in different oxidation levels of catechins. Green tea was found to possess the highest total phenolic content, and black tea showed the highest content of polyphenol oxidation products after the fermentation process [24].

Moreover, this study has enabled researches to optimize new conditions of HPLC for detection of several organic acids including glucuronic, gluconic, DSL, ascorbic, acetic, and succinic acid in kombucha tea. All organic acids were clearly separated by isocratic elution buffer using the C18 conventional column. Moreover, an elution buffer of 20 mM KH_2PO_4, pH 2.4, and a UV detector of 210 nm were suitable for the separation of each organic acid. Low molecular weight organic acids with high levels of polarity produced the greatest level of separation using mobile phase 20 mM KH_2PO_4 as an elution buffer. In addition, a low pH buffer was used to ensure that all acidic groups were protonated, which allowed for protection of the change from organic acids to the neutral form, thus allowing the best interaction between the organic acids and the C18 stationary phase [25]. The HPLC result revealed that gluconic acid and acetic acid were identified as the major organic acids in the kombucha tea that was prepared via the fermentation of green, oolong, and black tea for a period of 15 days.

In addition, kombucha prepared from black tea presented higher values of organic acid content in terms of glucuronic, gluconic, DSL, ascorbic, acetic, and succinic acid, when compared to oolong tea and green tea. The highest level of organic acid content in kombucha tea was found to be gluconic acid. The glucuronic acid present in kombucha has been associated with a number of benefits to the liver. Glucuronic acid plays an important role in liver detoxification and in the process associated with the excretion of exogenous chemicals known as glucuronidation [26].

The DPPH scavenging abilities of the kombucha prepared from green, oolong, and black teas significantly increased from the beginning of fermentation. During kombucha fermentation, many compounds with radical scavenging properties were obtained from the tea leaves and were considered by-products of the metabolic pathway of microorganisms. Catechins belong to polyphenols in green tea and they display high levels of antioxidant properties. Catechins also have the ability to scavenge free radicals and reactive oxygen species [13]. Notably, the increase of antioxidant potential against DPPH radicals from kombucha tea significantly reduced oxidative injuries in rats [27]. Moreover, kombucha tea prepared from green, oolong, and black teas showed significantly high amounts of total phenolic contents on day 3 of fermentation. After 3–6 days of kombucha tea fermentation, the amounts of phenolic compounds could maintain stability and then continued to be stable during 15 days of the fermentation process. The kinetics of microorganisms in kombucha fermentation increased around 3 days after beginning inoculation, which might be the reason for the enhancement of the phenolic compounds [28]. Many enzymes are produced during kombucha fermentation, such as phytase, α-galactosidase, and tannase, which are all related to the degradation of complex polyphenols to small molecules and are known to cause an increase in total phenolic compounds [29].

The study of kombucha prepared from Chinese black tea, Chinese green tea, Chinese oolong tea and Sri Lankan black tea also yielded the highest amounts of phenolic compounds on day 1 of fermentation. Notably, the phenolic compounds maintained a level of stability throughout 7 days of the kombucha fermentation period. Additionally, individual polyphenol contents yielded variations in terms of quantity in each type of tea [30]. The enzymes that were liberated from the bacteria and yeast in the tea fungus consortium degraded the complex of phenolic compounds in the tea, and the degradation was increased in the acidic environment of the fermentation process [21].

Moreover, kombucha contains other antioxidant substances, such as ascorbic acid and DSL, which were shown to be present in high levels in the kombucha that had been prepared from black tea. DSL, a derivative of D-glucaric acid, demonstrated detoxification, anticarcinogenic, and cholesterol-reduction properties [5,31–33]. DSL has been found to reduce ameliorate alloxan-induced type 1 diabetes by inhibiting the apoptotic death of pancreatic β-cells [5]. Moreover, DSL also revealed the greatest benefit in terms of anti-oxidative properties and displayed the ability to decrease oxidative damage to certain cellular biomolecules, such as on the lipids and proteins found in human blood platelets [34]. In addition, kombucha prepared from green tea revealed the highest level of phenolic content. An increase in the total phenolic content during kombucha fermentation that occurred from the ability of the bacteria and yeast was found to liberate enzymes, such as phytase, which could break down the cellulosic backbone of the tea leaves to release polyphenol compounds [21].

In this study, the kombucha prepared from green, oolong and black teas efficiently inhibited all tested pathogenic enteric bacteria: *Escherichia coli*, *E. coli* O157:H7 DMST 12743, *Shigella dysenteriae* DMST 1511, *Salmonella* Typhi DMST 22842, and *Vibrio cholerae*. The strongest antibacterial activity of kombucha tea was related to the presence of organic acid, such as the acetic acid found in kombucha tea. This antibacterial activity displayed a significant level of inhibitory activity against all tested bacteria. The studies by Dibner and Buttin [35] reported that organic acids displayed antimicrobial activity and also displayed an inhibitory effect against acid-intolerant species such as *E. coli*, *Salmonella* sp., and *Campylobacter* sp. that had been obtained from the guts of piglets. Weak organic acids, such as acetic acid and benzoic acid, were reported to show antimicrobial activity since the organic acid molecules could induce cytoplasmic acidification and destroy bacterial cells [36].

Kombucha tea displayed a remarkable level of antimicrobial activity against a broad range of microorganisms, which have demonstrated an ability to inhibit the growth of pathogens such as *Helicobacter pylori, Escherichia coli, Entamoeba cloacae, Pseudomonas aeruginosa, Staphylococcus aureus, S. epidermidis, Agrobacterium tumefaciens, Bacillus cereus, Aeromonas hydrophila, Salmonella typhimurium, S. enteritidis, Shigella sonnei, Leuconostoc monocytogenes, Yersinia enterocolitica, Campylobacter jejuni,* and *Candida albicans* [3,37]. Both acetic acid and catechins are known to inhibit a range of Gram-positive and Gram-negative microorganisms [2]. Therefore, kombucha tea has been recognized as containing active compound substances that could inhibit bacterial pathogens [23,38]. Battikh et al. [39] reported that kombucha prepared from green tea could inhibit *S. epidermidis, S. aureus, Micrococcus luteus, S. typhimurium, E. coli, Listeria monocytogenes*, and *P. aeruginosa* with diameters ranging from 12 to 22 mm, while kombucha prepared from black tea showed inhibition zones against these bacteria ranging from 10.5 to 19 mm. Furthermore, other studies showed that antibacterial activities did not exclusively from acetic acid or other organic acids such as citric acid, lactic acid, malic acid, and pyruvic acid [40] but possibly from other biologically active components such as bacteriocins, proteins, enzymes, and tea-derived phenolic compounds as well as tannins originally present in the tea broth that could have been involved as antimicrobial substances [37,38].

In contrast, unfermented tea did not display any antimicrobial activity against the tested microorganisms. This probably occurred because of the low concentrations of tea broth (1%, *w*/*v*) and polyphenol levels. Thus, unfermented tea did not display any inhibitory effects against the tested microorganisms [41]. In this study, neutralized kombucha did not display any inhibitory activity on all tested bacteria. Therefore, the use of the kombucha tea will provide natural organic acids for health benefits. The antibacterial activity of kombucha occurred from the acidity present in the kombucha tea. The same result was obtained from the previous studies of Cetojevic-Simin et al. [42], all of whom reported that neutralized kombucha prepared from black tea could not inhibit pathogenic bacteria.

In addition, heat-denatured kombucha tea prepared after boiling (100 °C, 20 min) and autoclaving (121.5 °C for 15 min) inhibited all tested pathogenic enteric bacteria. These results confirm that the antimicrobial components in the kombucha are thermostable. Moreover, heat-denatured kombucha tea that was achieved by boiling could significantly inhibit bacteria more effectively than the heat-denatured kombucha tea that had been autoclaved. The acidity of kombucha tea treated by boiling was maintained and the amount of total acidity was significantly higher than in the kombucha tea that was treated by autoclave. Moreover, the amount of total acidity present between heat-denatured kombucha tea by boiling and kombucha tea without any treatment was not significantly different. In contrast, the total acidity of kombucha tea that was treated by autoclave significantly decreased the amount of acidity when compared to kombucha tea without any treatment. This phenomenon indicated that the acidity activity of kombucha tea was reduced at high temperature by being autoclaved. This is particularly noteworthy with regard to the pasteurization of kombucha tea beverages that were prepared to preserve the potential antibacterial agents that are present in kombucha tea. Moreover, other components, such as catechins, and the antioxidant activities recorded after the autoclaving of these tea drinks at 120 °C for 20 min also decreased [43].

Kombucha prepared from green, oolong, and black tea were found to display effective toxicity against Caco-2 colorectal cancer cells. Interestingly, kombucha prepared from green tea and black tea showed the specific toxicity on cancer cells. Therefore, this is a new report about the cytotoxicity effect of kombucha prepared from different kinds of tea leaves on Caco-2 colorectal cancer cells. However, toxicity test of kombucha tea on different types of cancer and normal cell lines should be further studied.

The major beneficial components of green tea, such as EGCG, restrained carcinogenesis in a variety of tissues through the inhibition of mitogen-activated protein kinases, growth factor-related cell signaling, activation of activator protein 1 and nuclear factor-B, topoisomerase I, and matrix metalloproteinases along with other potential targets [44]. Previous reports have shown that kombucha prepared from black tea contained dimethyl 2-(2-hydroxy-2-methoxypropylidene) malonate and vitexin that caused certain cytotoxic effects on 786-O (human renal carcinoma) and U2OS (human osteosarcoma) cells by the reduction of cell invasion, cell motility, and matrix metalloproteinase activity [45,46]. Moreover, kombucha tea was found to significantly decrease the survival rate of prostate cancer cells by downregulating the expression of angiogenesis stimulators like matrix metalloproteinase, cyclooxygenase-2, interleukin-8, endothelial growth factor, and human inducible factor-1α [47]. The active substances of kombucha are associated with many of the compounds found in each type of the tea leaves and are related to the acid production that occurs from the microorganisms.

Tea polyphenols in kombucha that are present in the tea leaves and during kombucha fermentation were identified as anticancer substances. Tea polyphenols could inhibit the mechanisms of cancer formation such as gene mutation and cancer cell proliferation. Notably, tea polyphenols also induced the apoptosis of cancer cells and terminated cancer cell metastasis [48–50]. Both types of tea leaves that had been infused (1%, *w*/*v*) and acidified in the kombucha tea showed the ability to inhibit Caco-2 colorectal cancer cells. In 2013, a study by Zhao et al. [51] showed that fermented and unfermented specimens of Pu-erh tea and green tea could inhibit HT-29 colon cancer. Additionally, kombucha prepared from green tea and black tea could inhibit A549 human lung carcinoma cells and Hep-2 epidermoid carcinoma, while kombucha prepared from black tea could inhibit Hep-2 cells [23].

In this study, acetic acid in kombucha displayed toxicity against Caco-2 cancer cells. However, toxicity activities against cancer cells were also found to have occurred as a result of the presence of other organic acids such as glucuronic, gluconic, DSL, ascorbic, acetic, and succinic acid. DSL had the ability to inhibit the activity of glucuronidase enzymes, such as hydrolyzed glucuronides and produced aglycones. Aglycones are known to be toxic substances that are able to induce normal cells to become cancer cells [52]. Gluconic acid, glucuronic acid, lactic acid, and ascorbic acid are known to have the ability to reduce the occurrence of stomach cancer [53]. Notably, kombucha prepared from black tea was found to contain several organic acids. In this study, black tea revealed the lowest level of toxicity on Caco-2 cells because of the presence of other components found in the black tea, such as thearubigins and theaflavin, which were consistently degraded during kombucha fermentation. In contrast, catechins in green tea and oolong tea were not degraded during kombucha tea fermentation [22]. Theaflavins in black tea have been reported to possess activity against carcinogenesis by interfering with the signaling pathways and suppressing the transcription of certain oncoproteins [54]. Anticancer and antibacterial activity decreased when kombucha tea was neutralized at pH 7.0 by adjustments with NaOH. However, the anticancer components of kombucha tea were found to be thermostable after the tea was heated by either boiling or by being autoclaved. Especially, kombucha tea prepared from green and black tea after treatment by boiling at 100 °C for 20 min that also retained the cytotoxicity effect to cancer cells. However, kombucha tea at 15 days of fermentation revealed low pH values (2.70–2.94). Thus, kombucha tea should be prepared at a pH value of around 4.2 and should not be consume in amounts of more than 4 oz per day [55]. This scientific research study clarified that kombucha tea demonstrated the health benefits of effectively treating pathogenic enteric bacterial infection, anti-oxidation, and toxicity to colorectal cancer cells, which might help to promote the consumption of kombucha beverages among consumers.

5. Conclusions

In the present study, kombucha tea was fermented with a symbiotic culture of acetic acid bacteria and yeasts that produced a significant amount of organic acids. The kombucha tea prepared from different types of tea, namely green, oolong, and black tea, displayed significantly different values of pH, total acidity, total soluble solids, and organic acid content. Moreover, the HPLC system for the detection of several organic acids in kombucha tea was optimized in this study using the C18 conventional column. Isocratic elution buffer of 20 mM KH_2PO_4, pH 2.4 with 210 nm UV detector was used as a new adaptive HPLC condition for organic acid detection. Kombucha prepared from black tea at 15 days of fermentation showed the highest degree of organic acid content, such as with glucuronic, gluconic, DSL, ascorbic, acetic, and succinic acids. These organic acids were found to be effective against pathogenic enteric bacteria *Escherichia coli*, *E. coli* O157:H7, *Shigella dysenteriae*, *Salmonella* Typhi, and *Vibrio cholera*. They attributed to a large extent of the antibacterial effect observed to organic acids because when neutralizing the kombucha samples, antibacterial effect was not observed. Moreover, kombucha prepared from green tea and black tea demonstrated toxicity on Caco-2 colorectal cancer cells. These findings indicate the greatest potential health benefits of kombucha tea with regard to inhibiting pathogenic enteric bacteria and by promoting healthy function of the digestive system in the gastrointestinal tract. In addition, kombucha displayed antioxidant activity against DPPH radicals. Therefore, kombucha tea could be considered as a potential source of compounds presenting antioxidant activity, inhibitory activity against pathogenic enteric bacteria and selective toxicity on colorectal cancer cells.

Author Contributions: Conceptualization and Investigation, T.K., S.B., and Y.T.; Methodology, T.K., S.B., and Y.T.; Formal analysis, T.K. and Y.T.; Writing—original draft preparation, T.K. and Y.T.; Writing—review and editing, T.K. and Y.T.; Funding acquisition, Y.T.

Funding: This research study was supported by the Royal Golden Jubilee (RGJ) Ph.D. Program Scholarship, the Thailand Research Fund, grant no. PHD/0061/2558; and was partially supported by the Center of Excellence in Bioresources for Agriculture, Industry, and Medicine, Department of Biology, Faculty of Science, and The Graduate School, Chiang Mai University, Chiang Mai, Thailand.

Acknowledgments: This research work was partially supported by Chiang Mai University. The Tea Gallery Group (Thailand) Co., Ltd. and Chatchai Kitipornchai are also thoughtfully acknowledged. Department of Medical Technology, Faculty of Associated Medical Sciences, Department of Biology, Faculty of Science, and The Graduate School, Chiang Mai University, Chiang Mai, Thailand are also acknowledged.

Conflicts of Interest: The authors declare no conflict of interest.

References

1. Teoh, A.L.; Heard, G.; Cox, J. Yeast ecology of kombucha fermentation. *Int. J. Food Microbiol.* **2004**, *95*, 119–126. [CrossRef] [PubMed]
2. Jayabalan, R.; Malbasa, R.V.; Loncar, E.S.; Vitas, J.S.; Sathishkumar, M. A review on kombucha tea—Microbiology, composition, fermentation, beneficial effects, toxicity, and tea fungus. *Compr. Rev. Food Sci. Food Saf.* **2014**, *13*, 538–550. [CrossRef]
3. Dufresne, C.; Farnworth, E. Tea, Kombucha, and health. *Food Res. Int.* **2000**, *33*, 409–421. [CrossRef]
4. Banerjee, D.; Hassarajani, S.A.; Maity, B.; Narayan, G.; Bandyopadhya, S.K.; Chattopadhyay, S. Comparative healing property of kombucha tea and black tea against indomethacin induced gastric ulceration in mice: Possible mechanism of action. *J. Funct. Foods* **2010**, *1*, 284–293. [CrossRef]
5. Bhattacharya, S.; Gachhui, R.; Sil, P.C. The prophylactic role of D-saccharic acid-1,4-lactone against hyperglycemia-induced hepatic apoptosis via inhibition of both extrinsic and intrinsic pathways in diabetic rats. *J. Funct. Foods* **2013**, *4*, 283–296. [CrossRef]
6. Yang, Z.W.; Ji, B.P.; Zhou, F.; Li, B.; Luo, Y.; Yang, L.; Li, T. Hypocholesterolaemic and antioxidant effects of kombucha tea in high-cholesterol fed mice. *J. Sci. Food Agric.* **2009**, *89*, 150–156. [CrossRef]
7. Jigisha, A.; Nishant, R.; Navin, K. Green tea: A magical herb with miraculous outcomes. *Int. Res. J. Pharm.* **2012**, *3*, 139–148.
8. Peterson, J.; Dwyer, J.; Jacques, P.; Rand, W.; Prior, R.; Chui, K. Tea variety and brewing techniques influence flavonoid content of black tea. *J. Food Compos. Anal.* **2004**, *17*, 397–405. [CrossRef]
9. Cooper, R.; Morre, D.J.; Morre, D.M. Medicinal benefits of green tea. *J. Altern. Complement. Med.* **2005**, *11*, 521–528. [CrossRef]
10. Sajilata, M.G.; Poonam, R.B.; Singhal, R.S. Tea polyphenols as nutraceuticals. *Compr. Rev. Food Sci. Food Saf.* **2008**, *7*, 229–254. [CrossRef]
11. Chu, S.C.; Chen, C. Effects of origins and fermentation time on the antioxidant activities of kombucha. *Food Chem.* **2005**, *98*, 502–507. [CrossRef]
12. Talawat, S.; Ahantharik, P.; Laohawiwattanakul, S.; Premsuk, A.; Ratanapo, S. Efficacy of fermented teas in antibacterial activity. *Kasetsart J.* **2006**, *40*, 925–933.
13. Srihari, T.; Satyanarayana, U. Changes in free radical scavenging activity of kombucha during fermentation. *Int. J. Pharm. Sci. Res.* **2012**, *4*, 1978–1981.

14. Markov, S.L.; Malbasa, R.V.; Hauk, M.J.; Cvetkovic, D.D. Investigation of tea fungus microbe associations. The Yeasts. *Acta Period. Technol.* **2001**, *32*, 133–138.
15. Miliauskas, G.; Venskutonis, P.R.; van Beek, T.A. Screening of radical scavenging activity of some medicinal and aromatic plant extracts. *Food Chem.* **2004**, *85*, 231–237. [CrossRef]
16. Singhatong, S.; Leelarungrayub, D.; Chaiyasut, C. Antioxidant and toxicity activities of *Artocarpus lakoocha* Roxb. Heartwood extract. *J. Med. Plants Res.* **2010**, *4*, 947–953.
17. Battikh, H.; Chaieb, K.; Bakhrouf, A.; Ammar, E. Antibacterial and antifungal activities of black and green kombucha teas. *J. Food Biochem.* **2013**, *37*, 231–236. [CrossRef]
18. John, H.R.; Ferraro, M.J. *Methods for Dilution Antimicrobial Susceptibility Tests for Bacteria That Grow Aerobically*, 9th ed.; Clinical and Laboratory Standards Institute: Wayne, PA, USA, 2012; pp. 12–34.
19. Umthong, S.; Phuwapraisirisan, P.; Puthong, S. In vitro antiproliferative activity of partially purified *Trigona laeviceps* propolis from Thailand on human cancer cell lines. *BMC Complement. Altern. Med.* **2011**, *11*, 1–8. [CrossRef]
20. Ross, P.; Mayer, R.; Benziman, M. Cellulose biosynthesis and function in bacteria. *Microbiol. Rev.* **1991**, *55*, 35–58.
21. Blanc, P.J. Characterization of the tea fungus metabolites. *Biotechnol. Lett.* **1996**, *18*, 139–142. [CrossRef]
22. Bauer-Petrovska, B.; Petrushevska-Tozi, L. Mineral and water soluble vitamin content in the Kombucha drink. *Int. J. Food Sci. Technol.* **2000**, *35*, 201–205. [CrossRef]
23. Deghrigue, M.; Chriaa, J.; Battikh, H.; Abid, K.; Bakhrouf, A. Antiproliferative and antimicrobial activities of Kombucha tea. *Afr. J. Microbiol. Res.* **2013**, *7*, 3466–3470.
24. Jain, A.; Manghani, C.; Kohli, S.; Nigam, D.; Rani, V. Tea and human health. *Toxicol. Lett.* **2013**, *220*, 82–87. [CrossRef] [PubMed]
25. Horwitz, W. *Official Methods of Analysis of AOAC International*, 17th ed.; The Association of Official Analytical Chemists (AOAC International): Gaithersburg, MD, USA, 2000.
26. Ilmara, V.; Raimonds, L.; Arturs, P.; Pavels, S. Glucuronic acid from fermented beverages: Biomedical function in humans and its role in health protection. *IJRRAS* **2013**, *14*, 218–230.
27. Dipti, P.; Yogesh, B.; Kain, A.K.; Pauline, T.; Anju, B.; Sairam, M. Lead induced oxidative stress: Beneficial effects of kombucha tea. *Biomed. Environ. Sci.* **2003**, *16*, 276–282.
28. Jayabalan, R.; Marimuthu, S.; Swaminathan, K. Changes in content of organic acids and tea polyphenols during kombucha tea fermentation. *Food Chem.* **2007**, *102*, 392–398. [CrossRef]
29. Duenas, M.; Hernandez, T.; Estrella, I. Changes in the content of bioactive polyphenolic compounds of lentils by the action of exogenous enzymes. Effect on their antioxidant activity. *Food Chem.* **2007**, *101*, 90–97. [CrossRef]
30. Watawana, M.I.; Jayawardena, N.; Waisundara, V.Y. Value-added Tea (*Camellia sinesis*) as a functional food using the Kombucha 'Tea Fungus'. *Chiang Mai J. Sci.* **2018**, *45*, 136–146.
31. Hanausek, M.; Walasze, Z.; Slaga, T.J. Detoxifying cancer causing agents to prevent cancer. *Integr. Cancer Ther.* **2003**, *2*, 139–144. [CrossRef]
32. Olas, B.; Saluk-Juszczak, J.; Nowak, P.; Glowacki, R.; Bald, E.; Wachowicz, B. Protective effects of d-glucaro 1,4-lactone against oxidative/nitrative modifications of plasma proteins. *J. Nutr.* **2007**, *23*, 164–171. [CrossRef]
33. Walaszek, Z.; Szemraj, J.; Hanausek, M.; Adams, A.K.; Sherman, U. d-Glucaric acid content of various fruits and vegetables and cholesterol-lowering effects of dietary d-glucarate in the rat. *Nutr. Res.* **1996**, *16*, 673–681. [CrossRef]
34. Saluk-Juszczak, J.; Olas, B.; Nowak, P.; Staron, A.; Wachowicz, B. Protective effects of d-glucaro 1,4-lactone against oxidative modifications in blood platelets. *Nutr. Metab. Cardiovasc. Dis.* **2008**, *18*, 422–428. [CrossRef] [PubMed]
35. Dibner, J.J.; Buttin, P. Use of organic acids as a model to study the impact of gut microflora on nutrition and metabolism. *J. Appl. Poult. Res.* **2002**, *11*, 453–463. [CrossRef]
36. Ludovico, P.; Sansonetty, F.; Silva, M.T.; Corte-Real, M. Acetic acid induces a programmed cell death process in the food spoilage yeast *Zygosaccharomyces* bailii. *FEMS Microbiol. Lett.* **2003**, *3*, 91–96. [CrossRef]
37. Sreeramulu, G.; Zhu, Y.; Knol, W. Characterization of antimicrobial activity in kombucha fermentation. *Acta Biotechnol.* **2001**, *21*, 49–56. [CrossRef]
38. Sreeramulu, G.; Zhu, Y.; Knol, W. Kombucha fermentation and its antimicrobial activity. *J. Agric. Food Chem.* **2000**, *48*, 2589–2594. [CrossRef]

39. Battikh, H.; Bakhrouf, A.; Ammar, E. Antimicrobial effect of kombucha analogues. *J. Food Sci. Technol.* **2011**, *47*, 71–77. [CrossRef]
40. Hoffmann, N. The Ubiquitous Co-Enzyme UDP Glucuronic Acid Detoxifying Agent in Kombucha Tea. 2000. Available online: www.bluemarble.de/Norbert/kombucha/Glucuron/body_glucuron.htm (accessed on 1 July 2002).
41. Velicanski, A.S.; Cvetkovic, D.D.; Markov, S.L.; Tumbas, V.T.; Savatovic, S.M. Antimicrobial and antioxidant activity of lemon balm kombucha. *Acta Period. Technol.* **2007**, *38*, 165–172. [CrossRef]
42. Cetojevic-Simin, D.D.; Bogdanovic, G.M.; Cvetkovic, D.D.; Velicanski, A.S. Antiproliferative and antimicrobial activity of traditional Kombucha and *Satureja montana* L. Kombucha. *J. BUON* **2008**, *13*, 395–401.
43. Chen, Z.Y.; Zhu, Q.Y.; Tsang, D.; Huang, Y. Degradation of green tea catechins in tea drinks. *J. Agric. Food Chem.* **2001**, *49*, 477–482. [CrossRef]
44. Chen, L.; Zhang, H.Y. Cancer preventive mechanisms of the green tea polyphenol (-)-Epigallocatechin-3-gallate. *Molecules* **2007**, *12*, 946–957. [CrossRef] [PubMed]
45. Choi, H.J.; Eun, J.S.; Kim, B.G.; Kim, S.Y.; Jeon, H. Vitexin, an HIf-1α alpha inhibitor, has anti-metastatic potential in PC12 cells. *Mol. Cells* **2006**, *22*, 291–299. [PubMed]
46. Jayabalan, R.; Chen, P.N.; Hsieh, Y.S.; Prabhakaran, K.; Pitchai, P.; Marimuthu, S.; Thangaraj, P.; Swaminathan, K.; Yun, S.E. Effect of solvent fractions of kombucha tea on viability and invasiveness of cancer cells-characterization of dimethyl 2-(2-hydroxy-2-methoxypropylidine) malonate and vitexin. *Indian J. Biotechnol.* **2011**, *10*, 75–82.
47. Srihari, T.; Arunkumar, R.; Arunakaran, J.; Satyanarayana, U. Downregulation of signalling molecules involved in angiogenesis of prostate cancer cell line (PC-3) by kombucha (lyophilized). *Biomed. Prev. Nutr.* **2013**, *3*, 53–58. [CrossRef]
48. Conney, A.H.; Lu, Y.P.; Lou, Y.R.; Huang, M.T. Inhibitory effects of tea and caffeine on UV-induced carcinogenesis: Relationship to enhanced apoptosis and decreased tissue fat. *Eur. J. Cancer Prev.* **2002**, *2*, 28–36.
49. Ioannides, C.; Yoxall, V. Antimutagenic activity of tea: Role of polyphenols. *Curr. Opin. Clin. Nutr. Metab. Care* **2003**, *6*, 649–656. [CrossRef]
50. Park, A.M.; Dong, Z. Signal transduction pathways: Targets for green and black tea polyphenols. *J. Biochem. Mol. Biol.* **2003**, *6*, 66–77.
51. Zhao, X.; Song, J.L.; Kim, J.D.; Lee, J.S.; Park, K.Y. Fermented Pu-erh tea increases In Vitro anticancer activities in HT-29 cells and has antiangiogenetic effects on HUVECs. *J. Environ. Pathol. Toxicol. Oncol.* **2013**, *32*, 275–288. [CrossRef]
52. Wang, K.; Gan, X.; Tang, X.; Wang, S.; Tan, H. Determination of D-saccharic acid-1,4-lactone from brewed kombucha broth by high performance capillary electrophoresis. *J. Chromatogr. B* **2010**, *878*, 371–374. [CrossRef]
53. Hemila, H.; Herman, Z. Vitamin C and the common cold: A retrospective analysis of chalmers review. *J. Am. Coll. Nutr.* **1995**, *14*, 116–123. [CrossRef]
54. Weisburger, J.H.; Hara, Y.; Dolan, L.; Luo, F.Q.; Pittman, B.; Zang, E. Tea polyphenols as inhibitors of mutagenicity of major classes of carcinogens. *Mutat. Res.* **1996**, *371*, 57–63. [CrossRef]
55. Centers for Disease Control and Prevention, CDC. Unexplained severe illness possibly associated with consumption of kombucha tea—Iowa. *MMWR-Morb. Mortal. Wkly. Rep.* **1995**, *44*, 892–900.

Review

Silver Nanoparticles against Foodborne Bacteria. Effects at Intestinal Level and Health Limitations

Irene Zorraquín-Peña, Carolina Cueva, Begoña Bartolomé and M. Victoria Moreno-Arribas *

Institute of Food Science Research (CIAL), CSIC-UAM, C/Nicolás Cabrera 9, Campus de Cantoblanco, 28049 Madrid, Spain; irene.zorraquin@csic.es (I.Z.-P.); carolina.cueva@csic.es (C.C.); b.bartolome@csic.es (B.B.)
* Correspondence: victoria.moreno@csic.es; Tel.:+34-91-0017969; Fax: +34-910017905

Received: 5 December 2019; Accepted: 13 January 2020; Published: 17 January 2020

Abstract: Foodborne diseases are one of the factors that endanger the health of consumers, especially in people at risk of exclusion and in developing countries. The continuing search for effective antimicrobials to be used in the food industry has resulted in the emergence of nanotechnology in this area. Silver nanoparticles (Ag-NPs) are the nanomaterial with the best antimicrobial activity and therefore, with great potential of application in food processing and packing. However, possible health effects must be properly addressed to ensure food safety. This review presents a detailed description on the main applications of Ag-NPs as antimicrobial agents for food control, as well as the current legislation concerning these materials. Current knowledge about the impact of the dietary exposure to Ag-NPs in human health with special emphasis on the changes that nanoparticles undergo after passing through the gastrointestinal tract and how they alter the oral and gut microbiota, is also summarized. It is concluded that given their potential and wide properties against foodborne pathogens, research in Ag-NPs is of great interest but is not exempt from difficulties that must be resolved in order to certify the safety of their use.

Keywords: foodborne antimicrobials; silver nanoparticles; gut and microbiota; health

1. Introduction

It is estimated that there are 600 million cases of foodborne illnesses and 420,000 deaths annually worldwide. Unsafe foods are a risk to human health and countries' economy and mainly affect people at risk of exclusion, migrants and population under conflicts. The majority of foodborne diseases are related to pathogenic bacteria belonging to the genera *Salmonella*, *Listeria*, *Escherichia*, *Clostridium* and *Campylobacter*. Microbial contamination of food can occur at different stages of the process, such as harvesting, slaughtering, processing and distribution ("farm to fork") and can be caused by environmental contamination, such as water, soil or air [1]. The most common symptoms of foodborne diseases are gastrointestinal, such as diarrhea, but other consequences may be kidney and liver failure, brain and neural disorders, reactive arthritis and others. These diseases can be more severe in children, pregnant women, the elderly and those with a weakened immune system [2]. Traditional techniques such as salting, drying, freezing or fermentation are applied to extend the shelf life of food products, but there may be risk of recontamination. Therefore, there is a continuous need for antimicrobial agents that act in both food processing (preservation) and packaging (safety) stages [3].

In recent years, nanotechnology has experienced a noticeable rise in its applications, from agri-food to biotechnology, going through the engineering, cosmetic and textile industry. It can be considered a technological revolution [4]. Focusing on the field of food and health, nanotechnology is used in drug delivery system and nutrient release systems (nanoencapsulation), increasing the rate of recognition of disease symptoms and providing rapid treatments. It can also be applied to crops in the form of fertilizers and nanoscale additives or create nanoscale sensors to detect chemical, viral or bacterial contamination. In the case of food processing, it is a still emerging but promising technology [5].

Nanomaterials can be natural, accidental or manufactured and can be constituted by loose particles, aggregates or agglomerate in the form of nanoparticles, nanotubes, nanowires, nanofibers, and others. Of these, nanoparticles (NPs), wherein 50% or more of them in the numerical granulometry have one or more of the external dimensions between 1 and 100 nanometers, are possibly the most studied and the ones having more variety of sizes and shapes, which results in a large number of technological applications [6–8].

NPs are generally classified into organic and inorganic. Organic NPs incorporate carbon, whereas inorganic NPs incorporate metallic (Ag, Au, Cu), magnetic (Co, Fe, Ni), and/or semi-conductor components (ZnO, ZnS, CdS) [9]. Focusing our interest on silver nanoparticles (Ag-NPs), these have been widely used in medicine and biotechnology fields, due to their properties as antimicrobials. In this sense, numerous research studies have confirmed the effectiveness of Ag-NPs to inhibit the growth of pathogenic bacteria such as *Staphylococcus aureus*, *Streptococcus mutans*, *Streptococcus pyogenes*, *Escherichia coli* and *Proteus vulgaris* [10–12]. Interestingly, this activity has been also demonstrated using Ag-NPs obtained by 'biological methods' which are considered a great tool to reduce the negative effects associated with traditional nanoparticle synthesis commonly used in the laboratory [13]. In particular, two recent studies have shown the antimicrobial activity of Ag-NPs from apple pomace and from exopolysaccharides isolated from green microalgae against *E. coli* and *S. aureus* [14,15].

Shape, size, surface and charge are highlighted as the factors that influence the antimicrobial properties of Ag-NPs (Figure 1). Regarding shape (i.e., triangular, decahedron, spherical, cubic, platelet, among others), the spherical and the triangular forms seem to lead to higher antimicrobial activity [16–18]. Size is one of the most important factors when synthesizing nanoparticles, 1 to 30 nm being the most widely used range. Many studies have shown the size-dependent antibacterial activity of Ag-NPs [19–22]. Concerning the nanoparticle surface, it may be modified through the addition of coating agents, such as polymers (chitosan, polyethyleneimine, polyethylene glycol, polygamma glutamic acid), proteins (milk casein, bovine serum albumin, human serum albumin), antioxidants (glutathione) and/or polyvalent anion salts. Finally, the charge of the Ag-NPs determines their interaction with biological environments and its cellular uptake, which leads to a modulation of its antibacterial activity. Moreover, the antimicrobial activity of silver nanoparticles is bacteria strain- and cell wall structure-dependent [23].

The mechanisms of action by which Ag-NPs exert their antimicrobial effects are not completely clear, but two main hypothesis have been proposed: (*i*) a direct interaction of the nanoparticle with the cell membrane, and (*ii*) the release of ionic silver [24]. In the first hypothesis, the Ag-NPs would be adhered to the cell membrane via electrostatic attractions between the positive charges of the nanoparticles and the negative charges of the cells [25] or via the interaction of the nanoparticles into the sulfur and phosphorylated proteins present in the cell wall [26]. In any case, the interaction of the Ag-NPs with the cell membrane would produce its partial dissolution (Figure 1). In the second hypothesis, the Ag-NPs would enter into the cell and lead to a release of silver ions and the subsequent increase of reactive oxygen species (ROS) that would damage the enzymes involved in the cellular oxidation-reduction respiratory process and be finally responsible for cell death [16] (Figure 1). The two hypotheses could occur together as it has been showed that after interaction of the nanoparticle with the cell membrane, an internalization step takes place. In turn, this process can be affected by the nanoparticle charge [27]. Despite the antimicrobial effectiveness, some bacterial resistance against silver nanoparticles has been reported. Mechanisms such as negative regulation of porins, chromosomal resistance genes or plasmids with resistance genes have been proposed. However, this is still a field under study and more information to clarify this point at the frame of the food industry is clearly needed [24,28].

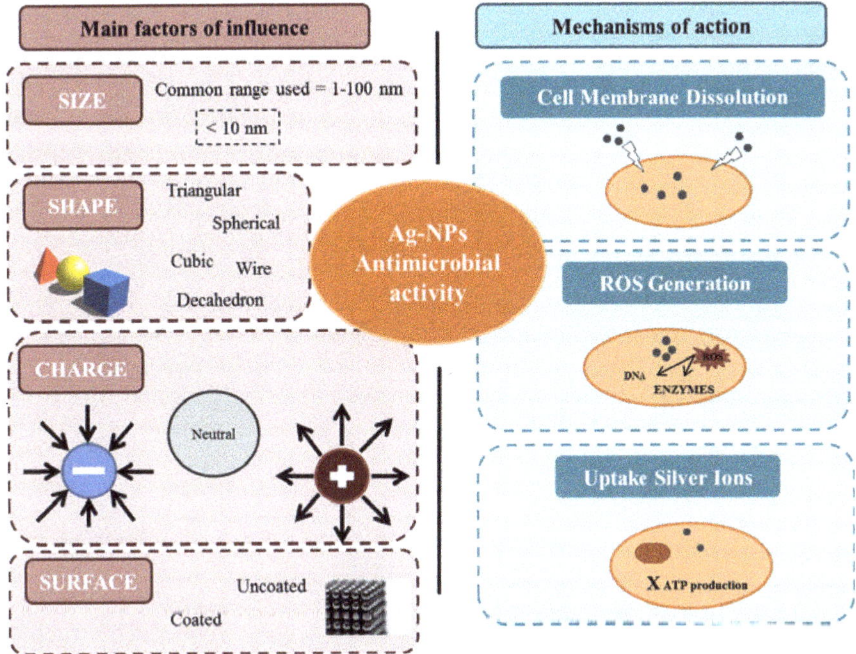

Figure 1. Main factors of influence and hypothetical mechanisms for the antimicrobial activity of silver nanoparticles.

On the other hand, the increased incorporation of silver nanoparticles into consumer products makes it essential to address their potential risk for human health. Nevertheless, there is still a lack of knowledge about their specific aspects of the intestinal uptake of silver nanoparticles [5]. The oral route of exposure has been poorly explored, despite the incorporation of such nanoparticles into packaging in contact with foods. After their ingestion, these nanoparticles pass through the digestive tract, where they may undergo physicochemical transformations, with consequences for the luminal environment, before crossing the epithelial barrier to reach the systemic compartment. Therefore, Ag-NPs toxicity and in particular, their effects at the gut level, are major concerns in the use and development of these nanomaterials.

This review presents a detailed description on the main applications of silver nanoparticles as antimicrobial agents for food control, as well as the current legislation concerning these materials. In addition, we summarize current knowledge about the impact of the dietary exposure to silver nanoparticles in human health with special emphasis in the gastrointestinal environment and microbiota, and highlight the areas where information is lacking. Finally, conclusions and future directions about both topics are summarized.

2. Applications of Antimicrobial Silver Nanoparticles in the Food Industry

Microbial food spoilage is a major global concern that can reduce the shelf life of food while increasing the risk of foodborne diseases. In this framework, the use of well-known potent antimicrobial agents such as silver nanoparticles constitutes an interesting approach. An overview of the effectiveness of silver nanoparticles to inhibit the growth of different microorganisms is shown in Table 1.

Table 1. Recompilation of studies about the antimicrobial effects of Ag-NPs against foodborne pathogens.

Ag-NPs Size	Ag-NPs Concentration	Gram (-) Pathogens	Gram (+) Pathogens	Yeast/Fungus	Main Results	Reference
-	0.034 µg Ag/mL	*Escherichia coli* K12	-	-	2 log reduction of *E. coli* after membrane filtration.	[29]
≈ 7 nm and 27.5 nm	0.26–26.5 mg Ag/dry g paper	*Escherichia coli*	*Enterococcus faecalis*	-	After filtration, the paper with a higher content of Ag-NPs almost completely deactivated bacterial growth. Reductions of 7 and 3 log were produced for *E. coli* and *E. faecalis*, respectively.	[30]
75 nm (spherical) and 8–20 nm (triangular)	-	*Escherichia coli, Pseudomonas aeruginosa, Salmonella typhi, Acinetobacter baumannii, Enterobacter cloacae, Haemophilus influenzae, Klebsiella pneumoniae, Neisseria mucosa, Proteus mirabilis, Serratia odorifera, Vibrio parahaemolyticus* and *Paenibacillus koreensis*	*Staphylococcus aureus, Bacillus subtilis* and *Paenibacillus koreensis*	-	The highest antimicrobial activity of the Ag-NPs was against *E. coli* and *P. aeruginosa*. For *S. typhi* and *B. subtilis* this activity was moderate and low for *S. aureus*.	[26]
14.6 nm	0.2, 0.5, 1, 1.5, 2 mg/mL	*Escherichia coli, Pseudomonas aeruginosa* and *Klebsiella pneumoniae*	*Lactobacillus rhamnosus* GG, *Bacillus cereus* and *Listeria monocytogenes*	*Aspergillus* and *Penicillium*	Inhibition of bacterial growth was dose dependent. *P. aeruginosa* was the bacteria most sensitive to Ag-NPs, followed by *E. coli*. On the contrary *L. monocytogenes* was the most resistant.	[31]
10, 20, 40, 60 and 80 nm	8 µg Ag/mL (10 nm), 11 µg Ag/mL (20 nm), 5 µg Ag/mL (40, 60 and 80 nm)	*Escherichia coli* and *Pseudomonas fluorescens*	-	*Saccharomyces cerevisiae*	Nanoparticles of a size equal to or less than 10 nm were more bioavailable when interacting with the cells. It was also shown that the toxicity of Ag-NPs decreased with increasing size.	[21]
8 nm (59 and 83 nm hydrodynamic size)	0–400 µg Ag/mL	*Proteus vulgaris* and *Shigella sonnei*	*Staphylococcus aureus, Bacillus megaterium*	-	The smaller size of Ag-NPs produced a greater growth inhibition. For both sizes the MIC values for the bacteria were between 75–400 ug/mL.	[20]
-	4.5 µg Ag/g film	*Pseudomonas* and *Enterobacteriaceae*	-	-	No significant differences were observed in the use of the film with nanoparticles compared to the conventional film.	[32]
10–50 nm	197 µg Ag/mL	*Campylobacter jejuni* (collection strain and isolates of patients and food chain)	-	-	The concentrations between 9.85 and 39.4 µg/mL were bactericidal after 24 h of incubation. In addition, the lower concentrations (1.23 and 4.92 µg/mL) significantly inhibited the growth of the collection strain.	[33]

Table 1. *Cont.*

Ag-NPs Size	Ag-NPs Concentration	Gram (-) Pathogens	Gram (+) Pathogens	Yeast/Fungus	Main Results	Reference
-	-	*Escherichia coli*	*Staphylococcus aureus*	*Aspergillus niger* and *Penicillium citrinum*	The antimicrobial activity of the chitosan, laponite and Ag-NPs hybrid film turned out to be slightly less than the chitosan film because laponite decreases the release of silver. There was also a greater inhibition of gram-positive bacteria compared to gram-negative bacteria.	[34]
20–30 nm	2.37, 4.75, 9.5 and 19 µg Ag/mL	*Escherichia coli* and *Salmonella typhimurium*	-	-	The concentration of 4.75 µg/mL Ag-NPs completely inhibited the growth of the two bacteria and the concentration of 9.5 µg/mL was sufficient to kill them.	[35]
47.3 nm	0–100 µg Ag/mL	*Escherichia coli* O157:H7, *Vibrio parahaemolyticu*, *Pseudomonas aeruginosa* and *Salmonella typhimurium*	*Listeria monocytogenes* and *Staphylococcus aureus*	-	Ag-NPs exerted a strong antimicrobial activity against all the pathogens tested. MIC of *V. parahaemolyticus* and *S. aureus* were 6.25 µg/mL and 50 µg/mL, respectively, and MBCs of *V. parahaemolyticus* and *S. aureus* were 12.5 µg/mL and 100 µg/mL, respectively.	[36]
6–25 nm (chemical synthesis) 80–120 nm and 40–100 nm (synthesized with *Fusarium nivale* and *Penicillium glabrum*)	170 µg Ag/mL	*Pseudomonas aeruginosa* PA01 4/4–15	*Bacillus cereus* B 504T UNIQEM, *Staphylococcus aureus* 209p	*Fusarium oxysporum*	Chemically synthesized AG-NPs inhibited microbial growth at 6 h of exposure, while with microbiologically synthesized nanoparticles it occurred at 24 h. *S. aureus* was the most resistant microorganism to both types of Ag-NPs.	[12]
5–15 nm	0.5, 1.0, 2.5, 5.0, 7.5, 10.0, 20.0 and 30.0 µg Ag/mL	*Escherichia coli*	*Staphylococcus aureus* and methicillin-resistant *Staphylococcus aureus*	-	The nanoparticles produced a total inhibition of *E. coli* growth at the concentration of 7.5 µg/mL. On the contrary, a concentration of >30 µg/mL is required for the complete inhibition of *S. aureus* and the resistant strain.	[15]
10–20 nm	8.34×10^{-7}, 3.61×10^{-6}, 5.79×10^{-5} and 4.63×10^{-4} mol/L	*Escherichia coli*	*Staphylococcus aureus*	-	Ag-NPs exerted a higher antimicrobial activity than the AgNO3 solution. This activity was concentration dependent and greater than other studies in which they use green synthesis due to their small size and spherical shape.	[14]

Table 1. *Cont.*

Ag-NPs Size	Ag-NPs Concentration	Gram (-) Pathogens	Gram (+) Pathogens	Yeast/Fungus	Main Results	Reference
-	-	*Salmonella typhimurium*	*Staphylococcus aureus*	-	The film that generated Ag-NPs in situ exerted a clear antimicrobial activity against both pathogens. A lower microbial growth was also observed when using this material to store chicken sausages for 4 days at 4 °C compared to the traditional film.	[37]
8–15 nm	30, 75, 150, and 300 µg Ag/mL	*Escherichia coli* O157:H7	*Listeria monocytogenes*	-	The material containing Ag-NPs exerted a greater antimicrobial activity against *E. coli* than against *L. monocytogenes* due to the greater wall thickness of the gram-positive bacteria.	[38]

Among other relevant results, Silvan et al. [33] demonstrated the antibacterial effect of Ag-NPs against multi-drug resistant (MDR) *Campylobacter* strains isolated from the chicken food chain and clinical patients. In another study, nanoparticles synthesized through *Forsythia suspensa* fruit water extract showed antibacterial activities against the most common foodborne pathogens, including *Listeria monocytogenes*, *Vibrio parahaemolyticus*, *Escherichia coli* O157:H7 and *Salmonella typhimurium* [36]. Similarly, the Ag-NPs synthesized from jack fruit seeds showed an antibacterial effect against *E. coli* and *S. typhimurium* [35]. The toxic effect of Ag-NPs synthesized using a bacterial exopolysaccharide as a reducing and stabilizing agent against various food pathogens (*L. monocytogenes*, *Aspergillus* spp. and *Penicillum* spp.) was also demonstrated [31]. Based on these promising results and in order to improve the shelf life and safety of food, there are various food preservation and safety strategies in which Ag-NPs have been employed (or proposed to be employed) in the food industry (Figure 2). Specifically, in this section, we report on the last studies assessing the use of silver nanoparticles in food processing and food packaging, and also the current regulation about it.

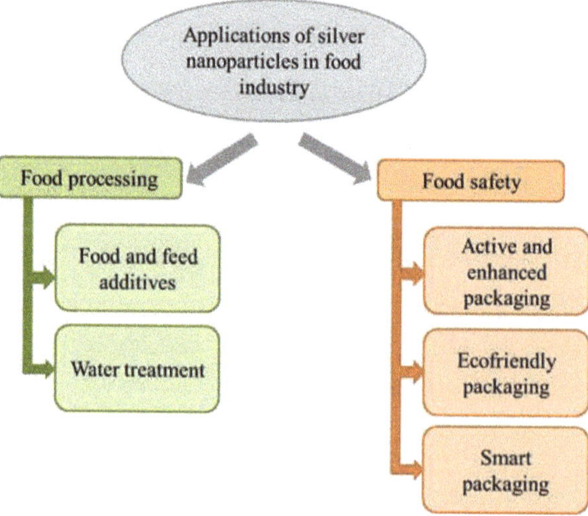

Figure 2. Classification of the use of Ag-NPs as antimicrobials in the food industry (adapted from Singh et al. [39]).

2.1. Food Processing (Preservation)

As feed additives, Ag-NPs has shown to be effective in the reduction of potentially pathogenic organisms such as *E. coli* and *Clostridium perfringens* [40–43], which could reduce the use of antibiotics in livestock [43]. Additionally, some Ag-NPs have also showed effective antiparasitic activity [44–46].

Ag-NPs have also been successfully applied in water treatment by incorporating them to filters with foam or by impregnating ultrafiltration membranes [29,30]. Although the investigation of Ag-NPs as a food additive is not very widespread, attractive attempts have been made to replace the use of sulfur dioxide by the use of antimicrobial nanoparticles. This is the case in the wine industry. For example, the effectiveness of a colloidal silver complex of a size < 1 nm was studied, managing to control the growth of lactic acid bacteria [47,48]. Another study confirmed the antimicrobial activity of two coated Ag-NPs against lactic acid bacteria and other microorganisms such as *S. aureus* and *E. coli*, with potential application in winemaking [49].

2.2. Food Packaging (Safety)

Food packaging is one of the areas where nanoparticles research and use is most relevant. The need of protection against foodborne diseases and the requirement of consumers to extend the useful life of the products urged the development of antimicrobial food packaging, special packaging that releases active biocide substances in order to improve the quality of the food [50]. The use of natural substances, such as green tea and chilto extracts and essential oils in packaging materials has already been investigated [51–53] but the use of Ag-NPs would be a more effective alternative because their antimicrobial activity is greater than phytochemicals. Nanotechnology in food packaging can be divided into three categories: (*i*) active packaging, (*ii*) ecofriendly packaging, and (*iii*) smart packaging, although packaging combinations are also possible (i.e., active and ecofriendly packing). In *active packaging*, the silver nanoparticles interact directly with the food or the environment polymer matrix which can be a non-degradable polymeric film such as polyethylene (PE), polyvinyl chloride (PVC) and ethylene vinyl alcohol (EVOH) or a biodegradable edible coating film made by a polymer or a stabilizing agent (ecofriendly packaging). In addition, Ag-NPs offer a good stability and slow release rates of silver ions in stored foods which makes them suitable candidates to be used in food packaging [54]. In line with this, Yu et al. [38] demonstrated the antibacterial effect of a material composed of Ag-NPs and cellulose nanofibrils against *E. coli* and *L. monocytogenes*. Similarly, silver nanoparticles immobilized with laponite showed a good growth inhibitory activity against *E. coli*, *S. aureus*, *A. niger* and *P. citrinum* [34]. Similar effects of silver nanoparticles against chicken meat (breasts and sausages) were found. The bacterial growth of *S. aureus*, *S. typhimurium* decreased, although there were also increases in cadaverine and thiamine [32,37]. On the other hand, the protective effect of silver nanoparticles in long-term packaging of nuts has also been demonstrated. The 3% silver package achieved a significant reduction in the presence of mold and coliforms and also achieved an antioxidant effect. Finally, the silver nanoparticles had a significant effect on increasing the shelf life of nuts [55].

At the framework of food safety, smart packaging, that is, packing with biosensors for the detection of pathogens represents a novel approach for food preservation, although still under development. The operation mode is based on the union or reaction of biological components with target species (microorganisms, toxins, etc.) and the transformation into detectable signals, which leads to the rapid detection of food contaminants [56]. Examples can be found in studies such as Abbaspour et al. [57] which described a selective sandwich biosensor for the detection of *S. aureus*. Combination of fluorescent carbon points (CDF) with silver nanoparticles has been reported for the detection and elimination of bacteria such as *E. coli* and *S. aureus* at low concentrations [58]. In the same way, the conjugated polyelectrolytes (CPs)–silver nanostructure pair has showed a high detection power against *E. coli* [59].

2.3. Regulation about Silver Nanoparticles Use in Foods and Food Industry Packaging (Safety)

The panel of the European Food Safety Agency (EFSA) on food additives and sources of nutrients added to food determined that there is insufficient information on Ag to assess its risk and, therefore, in the European Union (EU), Ag-NPs are not allowed in food supplements or food packaging unless authorized. EFSA has also provided Ag migration limits from the packaging (<0.05 mg/L in water and <0.05 mg/kg in food [60,61]. Therefore, manufacturers must carry out migration evaluations as well as genotoxicity, absorption, distribution, metabolism and *in vitro* excretion tests [60,61]. With all this information, EFSA will carry out a risk assessment of the specific case to determine if that package can be marketed or not. To date there are no known products that have been approved. On the other hand, in November 2015, Regulation (EU) 2015/2283 of the European Parliament and the European Council on new foods was approved. In this regulation, it appears the definition of "artificial nanomaterial" to include, within this new category ("novel foods"), all the foods that consist or contain artificial nanomaterials [62]. In spite of this, Ag-NPs still do not appear in the legislation of allowed food additives or in the materials in contact with food. Otherwise, in the United States, these regulations are influenced by the existing regulatory restrictions on the release of silver to the environment and are the responsibility of three agencies: the Environmental Protection Agency (EPA), the Food and Drug Administration (FDA) and the agency of the Institute National Occupational Safety and Health (NIOSH). The FDA published a guide for the use of nanotechnology in food or materials in contact with them and recommended that manufacturers study and prepare a toxicological profile for each container with nanomaterials [63,64].

As mentioned above, one of the problems of using these nanoparticles in food packaging is silver migration. Echegoyén and Nerín [65] conducted an analysis of the form of silver migration, whether ions or particles, into food simulants. They demonstrated that silver migrated to food and was dependent on food and warming, with acidic foods and oven heating presenting a higher migration. However, in their study, they found that Ag migration is well below the maximum migration limits established by European Union legislation. However, other studies did not observe any temperature or time-dependent increase in the migration of Ag packaged foods [66]. Gallocchio et al. [67] tested a container with Ag-NPs to store chicken breasts and did not observe that the silver content of the breasts was higher than that allowed by the EU.

3. Impact of Dietary Exposure to Silver Nanoparticles in Health: Gut Nanotoxicology Effects

As the investigation into the application of nanotechnology in the food sector increases, the potential of nanotechnology in food science/industry also expands and consequently, so does the human exposure to these substances. In the case of antimicrobial silver nanoparticles with application in food industry, the subject of this review, the main human exposure source is through the oral-gastrointestinal tract [68]. The mean dietary exposure level of Ag-NPs is estimated at 70–90 μg/day [69]. After ingestion, the Ag-NPs come in contact with lumen of the oral cavity and esophagus. There is little published information on the absorption rate of particulates through the epithelium of these two compartments, probably due to both a low surface area and a short residence time for most food matrices [68]. After that, during the gastrointestinal digestion process in the stomach and small intestine, the interaction of Ag-NPs with biological fluids can lead to its agglomeration, aggregation, and dissolution [69–73]. In addition, silver nanoparticle absorption (transcellular and paracellular transport and vesicular phagocytosis) through the gastrointestinal tract epithelium could take place. Finally, the nanoparticles that escape the absorption process reach the colon where they could modulate the composition and/or activity of gut microbiota, affecting the production and toxicity of bacterial metabolites [69]. Part of the initial intake of nanoparticles could be extracted in feces. According to the anatomy of the gastrointestinal tract, several environments characterizdc by specific microbiota composition are found. Gut microbiota harbors more than 100,000 billion microorganisms, including bacteria, fungi, viruses, protozoa and archaea, with bacteria representing a majority. The dominant gut bacterial phyla are the *Firmicutes* (including *Clostridium*, *Enterococcus*, *Lactobacillus*, and *Ruminococcus* genera) and

Bacteroidetes (including *Bacteroides* and *Prevotella* genera). These bacteria play an important role in the development and conservation of host health. Gut microbes play a role in human physiology through several mechanisms, including their contribution to nutrient and xenobiotic metabolism (e.g., synthesis of vitamins, digestion of oligo, and polysaccharides, drugs, etc.) and to the regulation of immune and neurodendocrine functions. Some of these effects are mediated by products of bacterial metabolism, such as short-chain fatty acids (SCFA), including propionate, butyrate or acetate, which influence the gut barrier, the inflammatory tone and the metabolic homeostatic control in different tissues [74]. To date, little is known about the effect of nanoparticles on the intestinal microbiota, but what is known is that there are numerous factors that can produce an imbalance in the intestinal bacterial populations, like food, triggering certain diseases. That is why the investigation of the NPs-gut microbiota relationship is so important and should continue [68,69].

The physical and chemical transformations of Ag-NPs during the gastrointestinal digestion could involve modifications in their toxic effect. Despite the specific features of these particles and the differences among them, they all display a close relationship between physicochemical reactivity and bioavailability/biopersistence in the gastrointestinal tract. Recently, Mercier-Bonin et al. [68] and Bouwmeester et al. [72] discussed the potential impact of the luminal and gastrointestinal environment on nanomaterial properties and toxicity studies. In this section, with a specific focus on silver nanoparticles, we report *in vitro* and *in vivo* studies considering both local and systemic levels effects, with a particular emphasis on their impact on gut microbiota.

3.1. *In Vitro Studies: Static and Dynamic Gut Simulators and Epithelium Cell Models*

Today, several *in vitro* models, from cell models to static and dynamic gastrointestinal models can be used alone or in combination for the study of Ag-NPs toxicity. As mentioned above, concentration/dose is a very important factor for the use of nanoparticles as an antimicrobial agent in the food field. In general, cytotoxicity of Ag-NPs is concentration-dependent. Moreover, depending on the cell type, silver nanoparticles cytotoxicity varies notably, and this should be taken into consideration for their application in consumer products [75]. As said above in relation to their antimicrobial activity, size, shape, charge and surface are also factors that affect the cytotoxicity of these nanoparticles. Ag-NPs' security depends on their state as they can form aggregates during their synthesis and use due to surface charge or they are covered by a high viscosity substance or suspended in a high viscosity environment. It has been shown that coated silver nanoparticles have lower cytotoxic due to the stabilization effect of the coating, which in turn, depends on the coating material and the thickness of the layer [76,77].

Different studies have evaluated the cytotoxic effect of silver nanoparticles in various human cell lines trying to understand the possible risks after exposure or ingestion (Table 2). However, today there are not many studies that evaluated the effect of silver nanoparticles in the oral cavity and the evaluation of the effect of these nanoparticles on oral microbiota is even more limited [68]. In one of these studies, it was found that Ag-NPs increased oxidative stress, inflammation and apoptosis in the human gingival fibroblast cell line (CRL-2014) [78]. Likewise, Niska et al. [79] observed that Ag-NP induced cell death in a concentration-dependent manner, not being toxic until concentrations greater than 40 μg/mL on human gingival fibroblasts (HGF-1). On the other hand, Hernández-Sierra et al. [80] studied the effect of Ag-NPs of different sizes of periodontal fibroblasts extracted from volunteers. They concluded that only nanoparticles with a size smaller than 20 nm increased the cytotoxicity of fibroblasts. Another study with human periodontal fibroblasts, specifically with the cell line HPLF, found that nanoparticles at low concentrations (≤16 μg/mL) had little influence on proliferation and cell cycle, while at high concentrations (32 and 64 μg/mL), they inhibited cell proliferation and significantly changed morphology [81]. The effect of Ag-NPs on oral bacteria has also been evaluated, with bacteria of the genus *Streptococcus* being more sensitive to them [82]. In another work, it was observed how the MIC and MBC of the silver nanoparticles was between 100 and 250 μg/mL for peri-implantitis pathogens [83]. On the other hand, Lu et al. [19] reported a MIC range between 25 and 50 μg/mL and this could be due to the smaller size of the nanoparticles used.

Table 2. Studies regarding silver nanoparticles cytotoxicity effects in several cell lines.

Cell Line	Ag-NPs Size	Main Results	Reference
Periodontal fibroblasts extracted from volunteers	<10 nm, 15–20 nm, and 80–100 nm	Small-sized Ag-NPs (<20 nm) increased cytotoxicity in cells in a dose and time dependent manner.	[80]
Human gingival fibroblast (CRL-2014)	2 nm	Ag-NPs increased oxidative stress, inflammation and cell apoptosis.	[78]
Human gingival fibroblasts (HGF-1)	10 nm	All the nanoparticles tested were less toxic and exerted a greater antimicrobial action than the silver nitrate solution.	[79]
Human periodontal fibroblasts (HPLF)	-	Ag-NPs at low concentration did not alter morphology or cell proliferation, while at high concentration they significantly altered morphology, inhibited proliferation, and stopped cell cycle.	[81]
Human colon epithelial cells (Caco-2)	-	There were no significant differences in cell viability between digested and undigested nanoparticles up to a concentration of 40 μg/mL. There was a viability reduction (65%) when adding a food matrix.	[84]
EpiIntestinal, EpiOral and EpiGinvival tissues	16 nm in average with sporadic occurrence of particles with a size of around 80 nm	Ag-NPs did not affect the viability of EpiOral and EpiGingival tissues. In addition, the release of IL-1 decreased significantly in EpiOral tissue. On the other hand, exposure of the EpiIntestinal tissue to gastric fluids with or without AG-NPs produced a slight decrease in viability.	[85]
Human colon epithelial cells (HT-28 and HCT-116)	6 nm	After 24 h of exposure with Ag-NPS, a decrease in dose-dependent cell viability was observed (2–10 μg/mL). A cytotoxicity of approximately 50% was reached at a concentration of 4 μg/mL.	[86]
Human colon epithelial cells (HT-29 and Caco-2) and colon regular cells (CCD-18)	10–50 nm	Cytotoxicity occurred in the cells at a concentration of Ag-NPs between 9.85 and 39.4 μg/mL.	[33]
Human colon epithelial cells (Caco-2)	≈7.74 nm	In this work, there was no significant decrease in cell viability after 24 h at a concentration of 100 μg/mL.	[87]
Human colon epithelial cells (Caco-2/HT-29-MTX)	51–52 nm	Cellular uptake decreased when using digested versus undigested Ag-NPs and the nanoparticles coated with lipolic acid dissolved to a greater extent than those coated with citrate.	[88]
Human colon epithelial cells (Caco-2)	5–25 nm for PEG-AgNPs 20; 4–6 nm and 10–50 nm for GSH-AgNPs	A significant decrease in cell viability was observed by exposing cells to digested nanoparticles (both coatings), but not to undigested nanoparticles.	[74]

Table 2. *Cont.*

Cell Line	Ag-NPs Size	Main Results	Reference
Rat brain microvessel endothelial cells (rBMEC)	25, 50 and 80 nm	Ag-NPs were more cytotoxic at lower concentrations for a size of 25 and 40 nm. On the contrary, for a size of 80 nm greater concentrations were needed.	[89]
Human breast epithelial cells (MCF-7)	20–80 nm	Ag-NPs caused apoptosis and necrosis in a dose-dependent manner to a concentration of 80 µg/mL. At higher concentrations, the apoptotic effect decreased while the necrotic effect became prominent.	[90]
Human liver epithelial cells (HepG2)	10 and 100 nm	Ag-NPs at low doses increased cell proliferation.	[91]
Human breast epithelial cells (MCF-7)	31.4 nm	Ag-NPs at a concentration of 60 µg/mL exhibited a cytotoxicity of 70% against the cell line. It was also observed that AgNP were much less cytotoxic when tested against a non-cancerous cell line.	[92]
Human dermal fibroblast (NHDF)	20–45 nm	Except for the sodium oleate and sodium dodecyl sulfate solutions, the rest prevented the aggregation of the nanoparticles, stabilized them and did not produce a significant cytotoxic effect on the cells.	[76]

Unlike what happens with the oral cavity, there are numerous *in vitro* investigations on the effect of silver nanoparticles in the intestine (Table 2). It was observed that the intake of Ag-NPs within a food matrix increased its absorption by colon epithelial cells, the opposite being the case when ingested without food. This shows us the ease with which nanoparticles can reach our intestines due to their consumption along with food [84]. The toxicity difference between digested and undigested silver nanoparticles was also studied. It was possible to verify how the undigested ones were mostly captured by the cellular model Caco-2/HT29-MTX [88]. In the study of Silvan et al. [33], exposure of GSH-Ag NPs to epithelial cells (HT-29, Caco-2 and CCD-18) showed a dose-dependent cytotoxic effect and no significant cytotoxicity occurred until concentrations of 4.93 µg/mL. This is supported by other works in which the toxicity of silver nanoparticles is usually in the range of 10 to 100 µg [93]. It was observed in the work of Vila et al. [87] that the exposure of small-sized Ag-NPs ($\approx$ 8 nm) at a concentration of 100 µg/mL only reached 20% cytotoxicity in Caco-2 cells. It was also shown that cell integrity was not altered using concentrations below 50 µg/mL.

The toxicity of these nanoparticles has not only been studied on oral and intestinal cell lines. There is a study in which non-cytotoxic doses of Ag-NPs were used against the HepG2 cell line. Moreover, at low doses (2 and 4 mg/L), Ag-NPs presented "hormesis" effects by accelerating cell proliferation and an activation of mitogen-activated protein kinase (MAPK) [91]. On the other hand, Khorrami et al. [92] described a cytotoxicity level of 70%, at concentrations between 10 and 60 µg/mL, on the MCF-7 breast cancer cell line, while for the L-929 cell line (non-carcinogenic), it was only 15%. In another study, the toxic effect of Ag-NPs on the MCF-7 cell line was also evaluated. Cellular cytotoxicity was observed from a nanoparticle concentration of 10 µg/mL [90]. This is opening the door to the use of this nanomaterial against cancer cells and therefore, to be a possible cancer therapy, alone or in combination with other existing methods [86,94]. Other studies reported that Ag-NPs may interact with the cerebral microvasculature producing a proinflammatory cascade in rat brain microvessel endothelial cells, as well as that larger NPs were less toxic, and blood–brain barrier (BBB) dysfunction and astrocyte swelling causing neuronal degeneration [89,95].

In reference to static models of gastrointestinal digestion (Table 3), there is one study that showed that Ag-NPs with a size of 60 nm and a concentration of 10 mg/mL (1661 particles/mL) in the presence of proteins survived the extreme conditions of the digestion and reached the intestine [71]. This probably means that epithelial cells of the intestine would be exposed to these nanoparticles, causing cellular damage. On the contrary, in the absence of proteins, the fraction of NPs that reached the intestine was smaller [71]. In other works they also studied the effect of nanoparticles during the passage through the gastrointestinal tract. It was found that by contacting them with synthetic human stomach fluid, the Ag-NPs aggregated significantly and also released ionic silver that was physically associated with the aggregates of particles such as silver chloride. In addition, it was seen that NPs smaller than 10 nm were added to a greater extent than larger one [96]. It was also demonstrated that depending on the composition and pH, the morphology and the size of the Ag-NPs changed when passing through the different fluids (simulated saliva and gastric and intestinal fluids); in addition, there was only a low toxicity in a pilot study of reconstituted human tissues model [85]. When Ag-NPs interact with proteins, a corona is always formed and it decreases the entry of nanoparticles into cells and therefore, cellular toxicity decreases [97]. Gil-Sánchez et al. [74] evaluated the effect of static *in vitro* digestion on silver nanoparticles with two types of coating. It was observed that the glutathione-coated nanoparticles agglomerated less than those that had the polyethylene glycol coating and were less toxic to colon cells. Studying the changes of NPs in dynamic models is more limited. In the work of Cueva et al. [98], the dynamic gastrointestinal simulator simgi® was used to digest Ag-NPs and study their effect on the colonic microbiota (Table 3). They did not observe changes in the bacterial composition or in the production of ammonium ions during the simulations, so it was concluded that Ag-NPs did not alter the composition and metabolic activity of the human intestinal microbiota. Another dynamic study showed that 90% of Ag-NPs were already dissolved by passing through the stomach and that many of the released ions bind to the food matrix. This results in less bioavailable ions and therefore, less toxicity (Table 3) [73].

Table 3. Studies in *in vitro* static and gastrointestinal simulation models regarding silver nanoparticles effects at gut level and microbiota.

Static/Dynamic	Particle Size	Main Results	Reference
Static	Ag-NPs 10–50 nm	The range of MIC and MBC for oral bacteria was between 100 and 250 µg/mL. Of the four oral bacteria tested, the most sensitive to silver nanoparticles were *Porphyromonas gingivalis* and *Fusobacterium nucleatum*.	[83]
Static	Ag-NPs 5, 15 and 55 nm	In this work it was observed that for the smaller nanoparticles the MIC was between 25 and 50 µg/mL. Oral aerobic bacteria were more susceptible than anaerobic bacteria.	[19]
Static	Ag-NPs 30–50 nm	A MIC between 15 and 90 µg/mL was reported for the exposure of Ag-NPs against 5 oral pathogens, much lower than for chlorhexidine.	[82]
Static	Ag-NPs 60 nm	AG-NPs of a size of 60 nm digested under physiological conditions can reach the wall of the intestine. It was also observed that after ingestion of Ag + ions nanoparticles ended up forming.	[71]

Table 3. *Cont.*

Static/Dynamic	Particle Size	Main Results	Reference
Static	Ag-NPs 10 and 75 nm	After the intake of Ag-NPs, these nanoparticles can be aggregated and chemically modified in the stomach depending on the size and surface chemistries.	[96]
Static	Ag-NPs 10 nm	There was a reduction in the production of capric and stearic fatty acids after exposure of the human feces sample to Ag-NPs, while palmitic acid increased. The presence of *Bacteroidetes* was also drastically reduced.	[99]
Static	16 nm in average with sporadic occurrence of particles with a size of around 80 nm	The size and morphology of the Ag-NPs changed due to the action of different gastric fluids and digestive enzymes. The study showed that nanoparticles agglomerate and partially react to form AgCl during exposure to fluids.	[85]
Static	Ag-NPs 14 nm	A decrease in *Bacteroidetes* and an increase in *Firmicutes* was observed, resulting in an alteration of the *Firmicutes/Bacteroidetes* ratio. Exposure with Ag-NPs for 24 h also altered the *Faecalibacterium prausnitzii* and *Clostridium coccoides/ Eubacterium rectal* taxa.	[100]
Static	5–25 nm for PEG-AgNPs 20; 4–6 nm and 10–50 nm for GSH-AgNPs	AgNPs agglomerated less and were less toxic in colon cells than PEG-AgNPs 20.	[74]
Dynamic	Ag-NPs 15 and 40 nm	It was observed that 90% of the silver nanoparticles had dissolved as they passed through the stomach and the resulting ions joined the digestive matrices.	[73]
Dynamic SIMulator of the GastroIntestinal tract (simgi®)	3–5 nm and 10–25 nm for PEG-AgNPs 20; 4–6 nm and 10–50 nm for GSH-AgNPs	Ingestion of Ag-NPs did not alter the microbial composition of the intestine or the metabolic activity of the bacteria. It was also observed how during the digestion the nanoparticle size was predominantly 3–5 nm, although small populations of agglomerates of these small nanoparticles were found.	[98]

A limited number of studies on the interaction of nanomaterials with the microbiome are available, most of them in rodents. In one *in vitro* study, it was observed that the Ag-NPs modified the *Firmicutes/Bacteroidetes* phylum ratio, increasing *Firmicutes* and decreasing *Bacteroidetes*. It was seen that the nanoparticles altered the intestinal microbiota as would a metabolic and inflammatory disease [100]. On the other hand, after exposure of silver nanoparticles (10 nm) to a concentration range of 0–100 μg/mL, a marked decrease in saturated fatty acids was observed, except in palmitic acid, which increased by 26–32%. The observation of these variations led to the sequencing of bacterial DNA. According to the results of Das et al. [99], Ag-NP ingestion, either deliberate or inadvertent, could have negative consequences on our intestinal microbiota, as evidenced by a significant decreasing of *Bacteroidetes* due to both ionic silver (AgCl; 25–200 mg/L) and nanosilver-mediated changes.

3.2. In Vivo Studies: Animal and Human Trials

When conducting studies *in vivo*, five main types of models have been used: rats, mice, *Caenorhabditis elegans*, fish (zebrafish), *Drosophila melanogaster* and, in a lesser extent, human studies (Table 4). Each model has its advantages and limitations, but all provide a great deal of information that helps us to conclude facts. Within all these models, rats and mice may be the most used, but the one that generates the most interest is the human model, since it provides real data when it comes to human applications.

Table 4. In vivo studies regarding silver nanoparticles effects at gut level and microbiota, organs and tissues.

Model	Study Design	Main Results	Reference
C57BL/6N mice	Ag-NPs 29,3 nm Dose: 100 mg/kg, 500 mg/kg or 1000 mg/kg	The production of significant alterations of selective genes in the caudate, frontal cortex and hippocampus of mice was observed after exposure to the nanoparticles. The data concluded that nanoparticles can produce neurotoxicity by generating oxidative stress.	[101]
Sprague–Dawley rats	Ag-NPs 60 nm; 28 days Four groups (10 rats in each group): vehicle control, low-dose group (30 mg/kg), middle-dose group (300 mg/kg), and high-dose group (1000 mg/kg)	A dose-dependent increased accumulation of Ag-NPs was observed in the lamina propria in both the small and large intestine, and also in the tip of the upper villi in the ileum and protruding surface of the fold in the colon. Rats that consumed nanoparticles also released more anormal mucus in the crypt lumen and ileal lumen and there was also detachment of cells at the tip of the villi.	[102]
F344 rats	Ag-NPs 56 nm; 13 weeks Four groups (10 rats in each group): vehicle control, low-dose (30 mg/kg), middle-dose (125 mg/kg), and high-dose (500 mg/kg).	Significant dose-dependent changes were found in alkaline phosphatase and cholesterol, indicating that exposure to more than 125 mg/kg of silver nanoparticles may result in slight liver damage. Histopathologic examination revealed a higher incidence of bile-duct hyperplasia, with or without necrosis. There was also a dose-dependent accumulation of silver in all tissues examined.	[103]
Mice	Ag-NPs 3–20 nm; 21 days Daily dose: 5, 10, 15 y 20 mg/kg	Mice treated with a dose of 10 mg/kg showed great weight loss. It was found that Ag-NPs damaged the microvilli of epithelial cells and intestinal glands. This may be the cause of weight loss due to intestinal malabsorption.	[104]
Wistar rats	Ag-NPs 10 nm; 14 days Daily dose: 0.02 mg/kg	Ag-NPs intake produced a synaptic degeneration and potential neuronal cell death due to alterations in synaptic structures and reduced levels of proteins associated with these structures	[105]
Sprague–Dawley rats	Ag-NPs 3–10 nm (98.7%), 10–30 nm (1.3%); 14 days Daily dose: 1 mg/kg or 10 mg/kg Three groups (6 rats in each group): control group, low-dose group (1 mg/kg), high-dose group (10 mg/kg)	After ingestion of Ag-NPs, neuron shrinkage, cytoplasmic or foot inflammation of the astrocytes and extravascular lymphocytes occurred. This led to the conclusion that Ag-NPs can induce neuronal degeneration and swelling of astrocytes even with oral exposure at low doses.	[106]

Table 4. *Cont.*

Model	Study Design	Main Results	Reference
C57BL/6NCrl mice	Ag-NPs 110 nm and 20 nm (PVP), 110 nm and 20 nm (Citrate); 28 days Daily dose: 10 mg/kg	None of the nanoparticles tested caused alterations in the structure or diversity of the intestinal microbiota of the mice.	[107]
Sprague–Dawley rats	Ag-NPs 10, 75 and 100 nm; 13 weeks Daily dose: 9, 18 and 36 mg/kg twice a day	It was possible to observe how the nanoparticles produced changes in the intestinal microbiota of the rats. There was an increase in Gram-negative bacteria. Exposure to smaller Ag-NPs resulted in a decrease in Lactobacillus spp. and the Firmicutes phyla.	[108]
Sprague–Dawley rats	Ag-NPs 12 nm; single exposure and multiple exposures over 30 days Daily doses: 2000 and 250 mg/kg for single and multiple administrations, respectively.	Single and multiple administrations resulted in silver accumulation in the liver, kidneys, spleen, stomach, and small intestine. But, concentrations of silver detected in tissues were far smaller than the administered doses (<99%), indicating its efficient excretion from the organism.	[109]
BALB/C mice	Ag-NPs 294 nm (NanoAg1) and 122 nm (NanoAg 2); 3 days Daily dose: 100 µL suspension	The administration of NanoAg1 increased the number of Clostridium perfringens and Escherichia coli and decreased that of Lactobacillus spp., But the results were not significant. NanoAg2 acted in reverse. It could also be seen how nanoparticle suspensions reversed a severe colonic lesion in mice.	[110]
Mice	Ag-NPs 55.17 nm; 28 days Doses: 0 (control), 11.4, 114 and 1140 µg Ag-NP/kg	In this work, an increase in the Firmicutes/Bacteroidetes ratio was observed, similar to that described in studies of obesity and inflammatory diseases.	[111]
Fish (*Piaractus mesopotamicus*)	Ag-NPs 50 nm; 24 h Dose: 0 (control), 2.5, 10, and 25 µg Ag-NPs/L	More silver accumulated in the brain than in gills and liver at all concentrations. There was also an increase in oxidative stress, as well as damage to the enterocytes in fish exposed to higher concentrations.	[112]
Zebrafish	Ag-NPs 58.6 nm; 14 days Dose: 500 mg/kg twice a day	Despite not finding lesions in the integrity of the intestinal epithelium, in this study it was observed that Ag-NPs decreased to a non-detectable level to beneficial bacterial populations of fish.	[113]
Zebrafish	Ag-NPs 10, 40 and 100 nm; 4 days Dose: 1, 5, 10, 50, 100, 150 y 200 ppm	It was observed that the salts and cations of the medium decreased the dissolution of the silver, thus limiting its action. Ag-NPs with a size of 10 and 100 nm caused developmental defects in the muscles and intestine of the embryo, while those of 40 nm produced lethal effects.	[114]
Zebrafish	Ag-NPs 20 and 100 nm; 96 h Dose: 0.61, 1.07, 0.67, and 1.28 mg/L	The coating of the nanoparticles increased the survival rate of the fish compared to the control. It was also observed that the smaller Ag-NPs were more lethal than the 100 nm. More nanoparticles accumulated in the intestines than in the gills.	[115]

Table 4. *Cont.*

Model	Study Design	Main Results	Reference
Caenorhabditis elegans	Ag-NPs 79 nm	The effect of silver nanoparticles for 10 generations of the nematode was studied. From the second a pronounced sensitization to the nanomaterial was observed.	[116]
Caenorhabditis elegans	Ag-NPs 25 and 75 nm; 12 h Dose: 5 mg/L	Exposure of E. coli to the nanoparticles and of the nematode to E. coli induced reproductive toxicity, as well as neurotoxicity.	[117]
Caenorhabditis elegans	Ag-NPs <100; 40 h Dose: 0, 1, 3, and 5 mg/kg	Different silver nanomaterials induce growth inhibition and reproductive toxicity when the soil is found at a concentration of ≥5 mg/kg.	[118]
Caenorhabditis elegans	Ag-NPs ≈ 69 nm	Factors that increased sensitivity and reproductive toxicity from the second generation could not be verified. Therefore, long-term risk cannot be assessed and other inheritance mechanisms, such as epigenetics, may be at play in multigenerational reproductive toxicity.	[119]
Drosophila melanogaster	Dose: 10–100 µg Ag/mL (accute intake) and 5 µg Ag/mL (chronic exposure)	After the acute intake, a significant toxic effect was observed at the concentration of 20 µg/mL and 50% of the flies could not complete their development cycle. In the case of the chronic exposure in 8 generations, a decrease in fertility was observed in the first three generations, after which it returned to normal.	[120]
Drosophila melanogaster	Ag-NPs 5–22 nm Dose: 10, 50, 100, 200 g/mL	All nanoparticles tested (synthesized from different natural extracts) significantly reduced the number of hatched larvae. In addition, those synthesized from mulberry, fig and olive produced a high mortality of larvae and adults.	[121]
Drosophila melanogaster	Ag-NPs 20–100 nm; 3, 10 and 30 days Dose: 5, 25, 50 and 250 µg Ag/mL	The effect of Ag-NPs depends on the dose and the stage of development of the flies. In general it alters the ability to lay eggs, decrease the size of the ovary and decrease survival and longevity.	[122]
Drosophila melanogaster	Ag-NPs 3.44 nm; 10 days Dose: 0.016, 0.08, 0.4, 1 y 2 mM	The 10 nM dose was completely toxic. Despite this, depigmentation was observed at all concentrations. Significant levels of intracellular ROS and DNA damage were also observed.	[123]
Humans	Volunteers: 60 Ag-NPs 5–10 nm (10 ppm) or 25–40 nm (32 ppm) Study 1: 10 ppm with 3, 7, and 14 day time periods Study 2: 32 ppm for 14 days Daily dose: 100 µg/day for 10 ppm, and 480 µg/day for 32 ppm	No significant changes were observed in metabolism, hematology, urine, physical findings, sputum morphology or changes in images. Nor were statistically significant changes detected in the markers of hydrogen peroxide production or peroxiredoxin protein expression. Instead, silver could be detected in human serum.	[124]

Regarding the findings with rat and mice, several studies have been carried out to evaluate the effect of these nanoparticles on the gastrointestinal tract. An abnormal mucus composition of the intestines of the animals was observed, as well as pigmentation of the villi and discharge of mucus granules [102,103,109]. In another study, it was discovered that Ag-NP damaged the microvilli of epithelial cells and intestinal glands in rats, thus decreasing the intestinal absorption of nutrients [104]. In the study by van den Brule et al. [111], by using Next Generation Sequencing (NGS), they observed how the intake of dietary doses Ag-NPs during 28 d did not significantly alter, in a dose-dependent manner, either the uniformity of the intestinal microbiota or populations in rats. But they could see an increase in the relationship between *Firmicutes* and *Bacteroidetes* phyla. Human and mouse gut microbiota are very similar at the phylum level, but not at the genera or species level; however, at least at the phylum level, these results could be extrapolated to humans. It was also discovered that the consumption of Ag-NPs modified the values of cholesterol and alkaline phosphatase in rats, which indicated that exposure to these nanoparticles could cause mild liver damage [103]. Silver nanoparticles are also easily able to cross the tight junction of the blood–brain barrier (BBB); therefore, they can be considered as neurotoxic. Rahman et al. [101] showed a neurotoxic effect induced by oxidative stress of Ag-NPs in three regions of the brain, including the caudate nucleus, the frontal cortex and the hippocampus of adult mice. In addition, another study showed that Ag-NPs produced neuronal degeneration and inflammation of astrocytes in the rat brain due to a low dose of exposure by oral and intragastric administration [105,106].

There are studies with other animal models like fishes. After exposure of fish at Ag-NPs concentrations of 2.5, 10, and 25 μg/L for 24 h, it was observed that the accumulation of silver in the brain was greater than in the liver and gills. In addition, fish that were exposed to the highest concentrations showed alterations in markers of oxidative stress [112]. In another study, various sizes of Ag-NPs coated with gum arabic (10, 40 and 100 nm) were used. Zebrafish embryos were exposed to various concentrations of these nanoparticles for 4 days and only an increase in lethality was observed with the 40 nm nanoparticles. This could be because of the retention of silver in the intestine depends on the particle size and the agglomerates [114]. In the same line is the work of Liu et al. [115], in which they demonstrated that the particle size is more influenced by the toxicity of Ag-NPs than the coating. Ag-NPs of small size (20 nm) and with citrate coating were more toxic and the toxic effect was greater in the intestine than in the gills or muscles. Merrifield et al. [113] showed, in adult zebrafish, that exposure to silver nanoparticles (500 mg/kg food) for 14 days had no effect on the richness and diversity of the microbiota. Similarly, Wilding et al. [107] found that the oral administration of silver nanoparticles of two different sizes (20 and 110 nm) and with two different coatings (PVP and citrate) for 28 days (10 mg/kg bw/day) did not change the diversity of the gut microbiome in mice. In another study, the effect of Ag-NPs in mouse models with inflammatory bowel disease was evaluated. A decrease in inflammation and a positive modulation of the gut microbiota could be observed [110]. By contrast, another study on rats fed twice-daily with oral silver nanoparticles for 13 weeks at various doses (9, 18 and 36 mg/kg bw/day) reported a general increase in the levels of Gram negative bacteria, and a decrease in the levels of *Firmicutes* [108]. It is important to note that there are differences between the human and zebrafish and rodent microbiome. Moreover, differences during gut transit and the interactions with the composition of the food matrix between animals and humans can affect nanomaterial properties in a different way during digestive transit and their putative effects.

There are also studies with *Caenorhabditis elegans*. In one of them, it was observed how the reactive oxygen species in the nematode increased when exposed to *E. coli* contaminated with Ag-NPs. They also increased reproductive toxicity and neurotoxicity [117]. Moon et al. [118] showed that the presence of different silver nanomaterials (including nanoparticles) in the soil decreased the growth and reproduction of *C. elegans*. Similarly, in another study, the hereditary reproductive toxicity produced by Ag-NPs in *C. elegans* was demonstrated and it was observed that this toxicity contributed to inducing germline mutations [116,119].

Finally, another of the most used non-human models is *Drosophila melanogaster*. In one of the studies, the larvae were fed with silver nanoparticles, which were able to reach the intestinal barrier. This was demonstrated by analyzing the increase in intracellular ROS [123]. Reproductive toxicity was also evaluated in this model. It was observed that exposure of adult specimens to Ag-NPs significantly affected the ability to lay eggs along with a deteriorated ovarian growth [121,122]. In a study of acute and chronic exposure, it was observed that the effect of a solution of Ag-NPs at a concentration of 20 ug/mL 50% of the larvae did not end their development cycle. In addition, after chronic exposure to an Ag-Nps solution of 5 ug/mL, it was shown that after three generations, the flies adapted to silver, recovering the fecundity lost in the first three [120].

As can be seen in the aforementioned paragraphs, Ag-NPs have been shown to have toxic effects to both in *in vitro* and *in vivo* models; however, there is a limited number of studies that reported the impacts of Ag-NPs on human health. One of them is the one carried out by Munger et al. [124]. A total of 60 healthy subjects ingested nanoparticles at concentrations of 10 and 32 ppm (Ag-NPs size: 5–10 nm) for 14 days. No significant changes were detected in the morphology of heart, lungs and other organs, nor in the reactive oxygen species or in the generation of proinflammatory cytokines. Nor did significant changes in metabolic measures appear in the conditions studied. The authors stressed the need to evaluate the effects of longer-term exposure.

Because of the increased potential for consumer exposure to Ag NP, it appeared urgent to assess the possible impact on the gut microbiota and on human health. As reviewed, few studies have investigated this issue and none are conclusive. The differences of results between studies could be related to the techniques used to analyze the microbiota. Moreover, it is difficult to make a comparison between studies published today because different sizes, shapes and concentrations of nanoparticles have been used. As a suggestion, future experiments should consider validated standards to ensure more comparable results and thus, make more reliable conclusions. Moreover, the transfer of results from animals to humans could be improved with the use of "humanized" animals by inoculation of human gut microbiota as well as by investigations conducted with longer exposure durations to better mimic human exposure scenarios.

4. Conclusions and Future Perspectives

Nanotechnology and specifically, silver nanoparticles, have a promising future ahead in the field of food. Silver nanoparticles have demonstrated extensive antimicrobial activity against foodborne pathogens as well as great effectiveness when they are incorporated into different types of packaging. Today, most studies focusing on the use of Ag-NPs in packaging are at the laboratory level and in most countries, are not allowed. In the European Union, in particular, more data are necessary to define the regulation of their employment. Therefore, investigation of the use of nanoparticles as a food additive is needed, as well as the evaluation of their effect on consumer health, since there are no long-term studies that assess the real concerns of their consumption. Very few studies have focused on the relationships between nanoparticles and oral microbiota, and, in the same way, effects of silver nanoparticles on the composition of the intestinal microbiota and the consequences on their metabolic activity are largely unknown. The range of models and diverse experimental conditions, such as *in vitro*, *ex vivo* and *in vivo* approaches, animal models and control conditions, make it even more difficult to compare the results and draw final conclusions. A crucial aspect for *in vitro* studies is to take care to incorporate the changing physiochemical properties of silver nanoparticles during transit of the gastrointestinal tract in the study design. It is also necessary to continue studying the different types of silver nanoparticles including form, size distribution as well as dose and modes of administration/exposure of them to state detrimental effects on health. Finally, the difficulties involved in the evaluation *in vivo* of the effects of ingested nanoparticles in the gut, due to differences between species (rodents vs. humans), may also be highlighted. Probable variability between individuals, not only in terms of the composition, but also in terms of the functional metabolic properties of the microbiota, should also be taken into account along with host physiological characteristics and environmental factors. In conclusion, given their potential and wide properties against foodborne pathogens, research into silver nanoparticles is of great interest for the food industry but is not exempt from difficulties that must be resolved in order to certify the safety of their use.

Author Contributions: Investigation and evaluation of published studies, I.Z.-P., C.C. and M.V.M.-A., Contributing in writing the manuscript, I.Z.-P., C.C., B.B. and M.V.M.-A., Designing of figures and tables, I.Z.-P.; C.C.; Supervision of the study, M.V.M.-A. All authors have read and agreed to the published version of the manuscript.

Funding: Research in our lab is funded by Grants AGL2015-64522-C2-R (Spanish Ministry of Economy and Competitiveness) and ALIBIRD-CM 2020 P2018/BAA-4343 (Comunidad de Madrid). I.Z.-P. thanks BES-2016-077980 contract.

Conflicts of Interest: The authors declare no conflict of interest.

References

1. WHO. Food Safety. 2019. Available online: https://www.who.int/news-room/fact-sheets/detail/food-safety (accessed on 1 October 2019).
2. Bari, M.L.; Yeasmin, S. Chapter 8—Foodborne Diseases and Responsible Agents. In *Food Safety and Preservation*; Grumezescu, A.M., Holban, A.M., Eds.; Academic Press: Cambridge, MA, USA, 2018; pp. 195–229.
3. Malhotra, B.; Keshwani, A.; Kharkwal, H. Antimicrobial food packaging: Potential and pitfalls. *Front. Microbiol.* **2015**, *6*, 611. [CrossRef] [PubMed]
4. Perez-Esteve, E.; Bernardos, A.; Martinez-Manez, R.; Barat, J.M. Nanotechnology in the development of novel functional foods or their package. An overview based in patent analysis. *Recent Pat. Food Nutr. Agric.* **2013**, *5*, 35–43. [CrossRef] [PubMed]
5. Monge, M.; Moreno-Arribas, M.V. Applications of Nanotechnology in Wine Production and Quality and Safety Control. In *Wine: Safety, Consumer Preferences, and Health*; Moreno-Arribas, M.V., Bartolomé, B., Eds.; Springer Life Sciences Publisher: New York, NY, USA, 2016; pp. 51–69.
6. Rao, C.N.R.; Cheetham, A.K. Science and technology of nanomaterials: Current status and future prospects. *J. Mater. Chem.* **2001**, *11*, 2887–2894. [CrossRef]
7. Edmundson, M.; Thanh, N.T.; Song, B. Nanoparticles based stem cell tracking in regenerative medicine. *Theranostics* **2013**, *3*, 573–582. [CrossRef] [PubMed]
8. Silva, L.P.; Silveira, A.P.; Bonatto, C.C.; Reis, I.G.; Milreu, P.V. Chapter 26—Silver Nanoparticles as Antimicrobial Agents: Past, Present, and Future. In *Nanostructures for Antimicrobial Therapy*; Ficai, A., Grumezescu, A.M., Eds.; Elsevier: Amsterdam, The Netherlands, 2017; pp. 577–596.
9. Rafique, M.; Sadaf, I.; Rafique, M.S.; Tahir, M.B. A review on green synthesis of silver nanoparticles and their applications. *Artif. Cells Nanomed. Biotechnol.* **2017**, *45*, 1272–1291. [CrossRef] [PubMed]
10. Abbaszadegan, A.; Ghahramani, Y.; Gholami, A.; Hemmateenejad, B.; Dorostkar, S.; Nabavizadeh, M.; Sharghi, H. The Effect of Charge at the Surface of Silver Nanoparticles on Antimicrobial Activity against Gram-Positive and Gram-Negative Bacteria: A Preliminary Study. *J. Nanomater.* **2015**, *2015*, 8. [CrossRef]
11. Jo, D.H.; Kim, J.H.; Lee, T.G.; Kim, J.H. Size, surface charge, and shape determine therapeutic effects of nanoparticles on brain and retinal diseases. *Nanomed. Nanotechnol. Biol. Med.* **2015**, *11*, 1603–1611. [CrossRef]
12. Gordienko, M.G.; Palchikova, V.V.; Kalenov, S.V.; Belov, A.A.; Lyasnikova, V.N.; Poberezhniy, D.Y.; Chibisova, A.V.; Sorokin, V.V.; Skladnev, D.A. Antimicrobial activity of silver salt and silver nanoparticles in different forms against microorganisms of different taxonomic groups. *J. Hazard. Mater.* **2019**, *378*, 120754. [CrossRef]
13. Singh, J.; Dutta, T.; Kim, K.H.; Rawat, M.; Samddar, P.; Kumar, P. 'Green' synthesis of metals and their oxide nanoparticles: Applications for environmental remediation. *J. Nanobiotechnol.* **2018**, *16*, 84. [CrossRef]
14. Ren, Y.Y.; Yang, H.; Wang, T.; Wang, C. Bio-synthesis of silver nanoparticles with antibacterial activity. *Mater. Chem. Phys.* **2019**, *235*, 121746. [CrossRef]
15. Navarro Gallón, S.M.; Alpaslan, E.; Wang, M.; Larese-Casanova, P.; Londoño, M.E.; Atehortúa, L.; Pavón, J.J.; Webster, T.J. Characterization and study of the antibacterial mechanisms of silver nanoparticles prepared with microalgal exopolysaccharides. *Mat. Sci. Eng. C Mater.* **2019**, *99*, 685–695. [CrossRef] [PubMed]
16. Pal, S.; Tak, Y.K.; Song, J.M. Does the antibacterial activity of silver nanoparticles depend on the shape of the nanoparticle? A study of the Gram-negative bacterium *Escherichia coli*. *Appl. Environ. Microbiol.* **2007**, *73*, 1712–1720. [CrossRef] [PubMed]
17. Sadeghi, B.; Garmaroudi, F.S.; Hashemi, M.; Nezhad, H.R.; Nasrollahi, A.; Ardalan, S.; Ardalan, S. Comparison of the anti-bacterial activity on the nanosilver shapes: Nanoparticles, nanorods and nanoplates. *Adv. Powder Technol.* **2012**, *23*, 22–26. [CrossRef]

18. Lu, W.; Yao, K.; Wang, J.; Yuan, J. Ionic liquids-water interfacial preparation of triangular Ag nanoplates and their shape-dependent antibacterial activity. *J. Colloid Interf. Sci.* **2015**, *437*, 35–41. [CrossRef] [PubMed]
19. Lu, Z.; Rong, K.; Li, J.; Yang, H.; Chen, R. Size-dependent antibacterial activities of silver nanoparticles against oral anaerobic pathogenic bacteria. *J. Mater. Sci. Mater. Med.* **2013**, *24*, 1465–1471. [CrossRef] [PubMed]
20. Khurana, C.; Vala, A.K.; Andhariya, N.; Pandey, O.P.; Chudasama, B. Antibacterial activity of silver: The role of hydrodynamic particle size at nanoscale. *J. Biomed. Mater. Res. A* **2014**, *102*, 3361–3368. [CrossRef] [PubMed]
21. Ivask, A.; Kurvet, I.; Kasemets, K.; Blinova, I.; Aruoja, V.; Suppi, S.; Vija, H.; Kakinen, A.; Titma, T.; Heinlaan, M.; et al. Size-dependent toxicity of silver nanoparticles to bacteria, yeast, algae, crustaceans and mammalian cells in vitro. *PLoS ONE* **2014**, *9*, e102108. [CrossRef]
22. Qing, Y.; Cheng, L.; Li, R.; Liu, G.; Zhang, Y.; Tang, X.; Wang, J.; Liu, H.; Qin, Y. Potential antibacterial mechanism of silver nanoparticles and the optimization of orthopedic implants by advanced modification technologies. *Int. J. Nanomed.* **2018**, *13*, 3311–3327. [CrossRef]
23. Duran, N.; Duran, M.; de Jesus, M.B.; Seabra, A.B.; Favaro, W.J.; Nakazato, G. Silver nanoparticles: A new view on mechanistic aspects on antimicrobial activity. *Nanomed. Nanotechnol.* **2016**, *12*, 789–799. [CrossRef]
24. Gugala, N.; Lemire, J.; Chatfield-Reed, K.; Yan, Y.; Chua, G.; Turner, R.J. Using a chemical genetic screen to enhance our understanding of the antibacterial properties of silver. *Genes* **2016**, *9*, 344. [CrossRef]
25. Dakal, T.C.; Kumar, A.; Majumdar, R.S.; Yadav, V. Mechanistic basis of antimicrobial actions of silver nanoparticles. *Front. Microbiol.* **2016**, *7*, 1831. [CrossRef] [PubMed]
26. Ghosh, S.; Patil, S.; Ahire, M.; Kitture, R.; Kale, S.; Pardesi, K.; Cameotra, S.S.; Bellare, J.; Dhavale, D.D.; Jabgunde, A.; et al. Synthesis of silver nanoparticles using Dioscorea bulbifera tuber extract and evaluation of its synergistic potential in combination with antimicrobial agents. *Int. J. Nanomed.* **2012**, *7*, 483–496.
27. Yue, Z.G.; Wei, W.; Lv, P.P.; Yue, H.; Wang, L.Y.; Su, Z.G.; Ma, G.H. Surface charge affects cellular uptake and intracellular trafficking of chitosan-based nanoparticles. *Biomacromolecules* **2011**, *12*, 2440–2446. [CrossRef] [PubMed]
28. Salas-Orozco, M.; Niño-Martínez, N.; Martínez-Castañón, G.; Torres Méndez, F.; Compean Jasso, M.E.; Ruiz, F. Mechanisms of resistance to silver nanoparticles in endodontic bacteria: A literature review. *J. Nanomater.* **2019**, *2019*, 11. [CrossRef]
29. Zodrow, K.; Brunet, L.; Mahendra, S.; Li, D.; Zhang, A.; Li, Q.; Alvarez, P.J. Polysulfone ultrafiltration membranes impregnated with silver nanoparticles show improved biofouling resistance and virus removal. *Water Res.* **2009**, *43*, 715–723. [CrossRef] [PubMed]
30. Dankovich, T.A.; Gray, D.G. Bactericidal paper impregnated with silver nanoparticles for point-of-use water treatment. *Environ. Sci. Technol.* **2011**, *45*, 1992–1998. [CrossRef] [PubMed]
31. Kanmani, P.; Lim, S.T. Synthesis and structural characterization of silver nanoparticles using bacterial exopolysaccharide and its antimicrobial activity against food and multidrug resistant pathogens. *Process Biochem.* **2013**, *48*, 1099–1106. [CrossRef]
32. Deus, D.; Kehrenberg, C.; Schaudien, D.; Klein, G.; Krischek, C. Effect of a nano-silver coating on the quality of fresh turkey meat during storage after modified atmosphere or vacuum packaging. *Poultr. Sci.* **2017**, *96*, 449–457. [CrossRef]
33. Silvan, J.M.; Zorraquin-Pena, I.; Gonzalez de Llano, D.; Moreno-Arribas, M.V.; Martinez-Rodriguez, A.J. Antibacterial activity of glutathione-stabilized silver nanoparticles against *Campylobacter* multidrug-resistant strains. *Front. Microbiol.* **2018**, *9*, 458. [CrossRef]
34. Wu, Z.; Huang, X.; Li, Y.C.; Xiao, H.; Wang, X. Novel chitosan films with laponite immobilized Ag nanoparticles for active food packaging. *Carbohydr. Polym.* **2018**, *199*, 210–218. [CrossRef]
35. Chandhru, M.; Logesh, R.; Rani, S.K.; Ahmed, N.; Vasimalai, N. One-pot green route synthesis of silver nanoparticles from jack fruit seeds and their antibacterial activities with escherichia coli and salmonella bacteria. *Biocatal. Agric. Biotechnol.* **2019**, *20*, 101241. [CrossRef]
36. Du, J.; Hu, Z.; Yu, Z.; Li, H.; Pan, J.; Zhao, D.; Bai, Y. Antibacterial activity of a novel Forsythia suspensa fruit mediated green silver nanoparticles against food-borne pathogens and mechanisms investigation. *Mater. Sci. Eng. C* **2019**, *102*, 247–253. [CrossRef] [PubMed]
37. Mathew, S.; Snigdha, S.; Mathew, J.; Radhakrishnan, E.K. Biodegradable and active nanocomposite pouches reinforced with silver nanoparticles for improved packaging of chicken sausages. *Food Packag. Shelf Life* **2019**, *19*, 155–166. [CrossRef]

38. Yu, Z.; Wang, W.; Kong, F.; Lin, M.; Mustapha, A. Cellulose nanofibril/silver nanoparticle composite as an active food packaging system and its toxicity to human colon cells. *Int. J. Biol. Macromol.* **2019**, *129*, 887–894. [CrossRef]
39. Singh, T.; Shukla, S.; Kumar, P.; Wahla, V.; Bajpai, V.K. Application of Nanotechnology in Food Science: Perception and Overview. *Front. Microbiol.* **2017**, *8*, 1501. [CrossRef]
40. Fondevila, M.; Herrer, R.; Casallas, M.C.; Abecia, L.; Ducha, J.J. Silver nanoparticles as a potential antimicrobial additive for weaned pigs. *Anim. Feed Sci. Tech.* **2009**, *150*, 259–269. [CrossRef]
41. Pineda, L.; Chwalibog, A.; Sawosz, E.; Lauridsen, C.; Engberg, R.; Elnif, J.; Hotowy, A.; Sawosz, F.; Gao, Y.; Ali, A.; et al. Effect of silver nanoparticles on growth performance, metabolism and microbial profile of broiler chickens. *Arch. Anim. Nutr.* **2012**, *66*, 416–429. [CrossRef]
42. Elkloub, K.; El Moustafa, M.E.; Ghazalah, A.A.; Rehan, A. Effect of dietary nanosilver on broiler performance. *Int. J. Poult. Sci.* **2015**, *14*, 177–182. [CrossRef]
43. Adegbeye, M.J.; Elghandour, M.M.M.Y.; Barbabosa-Pliego, A.; Monroy, J.C.; Mellado, M.; Ravi Kanth Reddy, P.; Salem, A.Z.M. Nanoparticles in Equine Nutrition: Mechanism of Action and Application as Feed Additives. *J. Equine Vet. Sci.* **2019**, *78*, 29–37. [CrossRef]
44. Dalloul, R.A.; Lillehoj, H.S. Poultry coccidiosis: Recent advancements in control measures and vaccine development. *Expert Rev. Vaccines* **2006**, *5*, 143–163. [CrossRef]
45. Chauke, N.; Siebrits, F.K. Evaluation of silver nanoparticles as a possible coccidiostat in broiler production. *S. Afr. J. Anim. Sci.* **2012**, *42*, 493–497. [CrossRef]
46. Gherbawy, Y.A.; Shalaby, I.M.; El-Sadek, M.S.A.; Elhariry, H.M.; Abdelilah, B.A. The anti-fasciolasis properties of silver nanoparticles produced by Trichoderma harzianum and their improvement of the anti-fasciolasis drug triclabendazole. *Int. J. Mol. Sci.* **2013**, *14*, 21887–21898. [CrossRef] [PubMed]
47. Izquierdo-Cañas, P.M.; García-Romero, E.; Huertas-Nebreda, B.; Gómez-Alonso, S. Colloidal silver complex as an alternative to sulphur dioxide in winemaking. *Food Control* **2012**, *23*, 73–81. [CrossRef]
48. Garde-Cerdán, T.; López, R.; Garijo, P.; González-Arenzana, L.; Gutiérrez, A.R.; López-Alfaro, I.; Santamaría, P. Application of colloidal silver versus sulfur dioxide during vinification and storage of Tempranillo red wines. *Aust. J. Grape Wine Res.* **2014**, *20*, 51–61. [CrossRef]
49. García-Ruiz, A.; Crespo, J.; López-de-Luzuriaga, J.M.; Olmos, M.E.; Monge, M.; Rodríguez-Álfaro, M.P.; Martín-Álvarez, P.J.; Bartolome, B.; Moreno-Arribas, M.V. Novel biocompatible silver nanoparticles for controlling the growth of lactic acid bacteria and acetic acid bacteria in wines. *Food Control* **2015**, *50*, 613–619. [CrossRef]
50. Carbone, M.; Donia, D.T.; Sabbatella, G.; Antiochia, R. Silver nanoparticles in polymeric matrices for fresh food packaging. *J. King Saud Univ. Sci.* **2016**, *28*, 273–279. [CrossRef]
51. Manso, S.; Cacho-Nerin, F.; Becerril, R.; Nerín, C. Combined analytical and microbiological tools to study the effect on Aspergillus flavus of cinnamon essential oil contained in food packaging. *Food Control* **2013**, *30*, 370–378. [CrossRef]
52. Medina-Jaramillo, C.; Ochoa-Yepes, O.; Bernal, C.; Famá, L. Active and smart biodegradable packaging based on starch and natural extracts. *Carbohydr. Polym.* **2017**, *176*, 187–194. [CrossRef]
53. Moreno, M.A.; Orqueda, M.E.; Gómez-Mascaraque, L.G.; Isla, M.I.; López-Rubio, A. Crosslinked electrospun zein-based food packaging coatings containing bioactive chilto fruit extracts. *Food Hydrocoll.* **2019**, *95*, 496–505. [CrossRef]
54. Duncan, T.V. Applications of nanotechnology in food packaging and food safety: Barrier materials, antimicrobials and sensors. *J. Colloid Interface Sci.* **2011**, *363*, 1–24. [CrossRef]
55. Tavakoli, H.; Rastegar, H.; Taherian, M.; Samadi, M.; Rostami, H. The effect of nano-silver packaging in increasing the shelf life of nuts: An in vitro model. *Ital. J. Food Saf.* **2017**, *6*, 6874. [CrossRef] [PubMed]
56. Inbaraj, B.S.; Chen, B.H. Nanomaterial-based sensors for detection of foodborne bacterial pathogens and toxins as well as pork adulteration in meat products. *J. Food Drug Anal.* **2016**, *24*, 15–28. [CrossRef] [PubMed]
57. Abbaspour, A.; Norouz-Sarvestani, F.; Noori, A.; Soltani, N. Aptamer-conjugated silver nanoparticles for electrochemical dual-aptamer-based sandwich detection of staphylococcus aureus. *Biosens. Bioelectron.* **2015**, *68*, 149–155. [CrossRef] [PubMed]
58. Roh, S.G.; Robby, A.I.; Phuong, P.T.M.; In, I.; Park, S.Y. Photoluminescence-tunable fluorescent carbon dots-deposited silver nanoparticle for detection and killing of bacteria. *Mater. Sci. Eng. C* **2019**, *97*, 613–623. [CrossRef] [PubMed]

59. Wang, X.; Cui, Q.; Yao, C.; Li, S.; Zhang, P.; Sun, H.; Lv, F.; Liu, L.; Li, L.; Wang, S. Conjugated Polyelectrolyte-Silver Nanostructure Pair for Detection and Killing of Bacteria. *Adv. Mater. Technol.* **2017**, *2*, 1700033. [CrossRef]
60. European Food Safety Authority (EFSA). Scientific opinion on the re-evaluation of silver (E 174) as food additive. *EFSA J.* **2016**, *14*, 4664.
61. European Food Safety Authority (EFSA). Guidance on risk assessment of the application of nanoscience and nanotechnologies in the food and feed chain: Part 1, human and animal health. *EFSA J.* **2018**, *16*, 5327.
62. Ávalos Fúnez, A.; Haza, A.; Morales, P. Nanotecnología en la industria alimentaria I: Aplicaciones. *Rev. Complut. Cienc. Vet.* **2016**, *10*, 1–17. [CrossRef]
63. Food and Drug Administration (FDA). *Considering Whether an FDA-Regulated Product Involves the Application of Nanotechnology*; FDA: Washington, DC, USA, 2014.
64. Richa, S.; Dimple, S.C. Regulatory Approval of Silver Nanoparticles. *Appl. Clin. Res. Clin. Trials Regul. Aff.* **2018**, *5*, 74–79.
65. Echegoyen, Y.; Nerín, C. Nanoparticle release from nano-silver antimicrobial food containers. *Food Chem. Toxicol.* **2013**, *62*, 16–22. [CrossRef]
66. Cushen, M.; Kerry, J.; Morris, M.; Cruz-Romero, M.; Cummins, E. Evaluation and Simulation of Silver and Copper Nanoparticle Migration from Polyethylene Nanocomposites to Food and an Associated Exposure Assessment. *J. Agric. Food Chem.* **2014**, *62*, 1403–1411. [CrossRef] [PubMed]
67. Gallocchio, F.; Cibin, V.; Biancotto, G.; Roccato, A.; Muzzolon, O.; Carmen, L.; Simone, B.; Manodori, L.; Fabrizi, A.; Patuzzi, I.; et al. Testing nano-silver food packaging to evaluate silver migration and food spoilage bacteria on chicken meat. *Food Addit. Contam. Part A Chem. Anal. Control Expo. Risk Assess.* **2016**, *33*, 1063–1071. [CrossRef] [PubMed]
68. Mercier-Bonin, M.; Despax, B.; Raynaud, P.; Houdeau, E.; Thomas, M. Mucus and microbiota as emerging players in gut nanotoxicology: The example of dietary silver and titanium dioxide nanoparticles. *Crit. Rev. Food Sci. Nutr.* **2018**, *8*, 1023–1032. [CrossRef] [PubMed]
69. Li, J.; Tang, M.; Xue, Y. Review of the effects of silver nanoparticle exposure on gut bacteria. *J. Appl. Toxicol.* **2019**, *39*, 27–37. [CrossRef] [PubMed]
70. Walczak, A.P.; Fokkink, R.; Peters, R.; Tromp, P.; Herrera Rivera, Z.E.; Rietjens, I.M.; Hendriksen, P.J.; Bouwmeester, H. Behaviour of silver nanoparticles and silver ions in an in vitro human gastrointestinal digestion model. *Nanotoxicology* **2013**, *7*, 1198–1210. [CrossRef]
71. Bouwmeester, H.; van der Zande, M.; Jepson, M.A. Effects of food-borne nanomaterials on gastrointestinal tissues and microbiota. *WIREs Nanomed. Nanobiotechnol.* **2018**, *10*, e1481. [CrossRef]
72. Bove, P.; Malvindi, M.A.; Kote, S.S.; Bertorelli, R.; Summa, M.; Sabella, S. Dissolution test for risk assessment of nanoparticles: A pilot study. *Nanoscale* **2017**, *9*, 6315–6326. [CrossRef]
73. Gil-Sánchez, I.; Monge, M.; Miralles, B.; Armentia, G.; Cueva, C.; Crespo, J.; de Luzuriaga, J.M.L.; Olmos, M.E.; Bartolomé, B.; de Llano, D.G.; et al. Some new findings on the potential use of biocompatible silver nanoparticles in winemaking. *Innov. Food Sci. Emerg.* **2019**, *51*, 64–72. [CrossRef]
74. Cueva, C.; Gil-Sánchez, I.; Ayuda-Durán, B.; González-Manzano, S.; González-Paramás, A.M.; Santos-Buelga, C.; Moreno-Arribas, M. An integrated view of the effects of wine polyphenols and their relevant metabolites on gut and host health. *Molecules* **2017**, *22*, 99. [CrossRef]
75. Akter, M.; Sikder, M.T.; Rahman, M.M.; Ullah, A.K.M.A.; Hossain, K.F.B.; Banik, S.; Hosokawa, T.; Saito, T.; Kurasaki, M. A systematic review on silver nanoparticles-induced cytotoxicity: Physicochemical properties and perspectives. *J. Adv. Res.* **2018**, *9*, 1–16. [CrossRef]
76. Verkhovskii, R.; Kozlova, A.; Atkin, V.; Kamyshinsky, R.; Shulgina, T.; Nechaeva, O. Physical properties and cytotoxicity of silver nanoparticles under different polymeric stabilizers. *Heliyon* **2019**, *5*, e01305. [CrossRef] [PubMed]
77. Fahmy, H.M.; Mosleh, A.M.; Elghany, A.A.; Shams-Eldin, E.; Abu Serea, E.S.; Ali, S.A.; Shalan, A.E. Coated silver nanoparticles: Synthesis, cytotoxicity, and optical properties. *RSC Adv.* **2019**, *9*, 20118–20136. [CrossRef]
78. Inkielewicz-Stepniak, I.; Santos-Martinez, M.J.; Medina, C.; Radomski, M.W. Pharmacological and toxicological effects of co-exposure of human gingival fibroblasts to silver nanoparticles and sodium fluoride. *Int. J. Nanomed.* **2014**, *9*, 1677–1687.

79. Niska, K.; Knap, N.; Kedzia, A.; Jaskiewicz, M.; Kamysz, W.; Inkielewicz-Stepniak, I. Capping Agent-Dependent Toxicity and Antimicrobial Activity of Silver Nanoparticles: An In Vitro Study. Concerns about Potential Application in Dental Practice. *Int. J. Med. Sci.* **2016**, *13*, 772–782. [CrossRef]
80. Hernandez-Sierra, J.F.; Galicia-Cruz, O.; Angelica, S.A.; Ruiz, F.; Pierdant-Perez, M.; Pozos-Guillen, A.J. In vitro cytotoxicity of silver nanoparticles on human periodontal fibroblasts. *J. Clin. Pediatr. Dent.* **2011**, *36*, 37–41. [CrossRef]
81. Tang, X.; Li, L.; Meng, X.; Liu, T.; Hu, Q.; Miao, L. Cytotoxicity of Silver Nanoparticles on Human Periodontal Ligament Fibroblasts. *Nanosci. Nanotechnol. Lett.* **2017**, *9*, 1015–1022. [CrossRef]
82. Panpaliya, N.P.; Dahake, P.T.; Kale, Y.J.; Dadpe, M.V.; Kendre, S.B.; Siddiqi, A.G.; Maggavi, U.R. In vitro evaluation of antimicrobial property of silver nanoparticles and chlorhexidine against five different oral pathogenic bacteria. *Saudi Dent. J.* **2019**, *31*, 76–83. [CrossRef]
83. Vargas-Reus, M.A.; Memarzadeh, K.; Huang, J.; Ren, G.G.; Allaker, R.P. Antimicrobial activity of nanoparticulate metal oxides against peri-implantitis pathogens. *Int. J. Antimicrob. Agents* **2012**, *40*, 135–139. [CrossRef]
84. Lichtenstein, D.; Ebmeyer, J.; Knappe, P.; Juling, S.; Bohmert, L.; Selve, S.; Niemann, B.; Braeuning, A.; Thunemann, A.F.; Lampen, A. Impact of food components during in vitro digestion of silver nanoparticles on cellular uptake and cytotoxicity in intestinal cells. *Biol. Chem.* **2015**, *396*, 1255–1264. [CrossRef]
85. Pind'áková, L.; Kašpárková, V.; Kejlová, K.; Dvořáková, M.; Krsek, D.; Jírová, D.; Kašparová, L. Behaviour of silver nanoparticles in simulated saliva and gastrointestinal fluids. *Int. J. Pharm.* **2017**, *527*, 12–20. [CrossRef]
86. Gurunathan, S.; Qasim, M.; Park, C.; Yoo, H.; Kim, J.H.; Hong, K. Cytotoxic Potential and Molecular Pathway Analysis of Silver Nanoparticles in Human Colon Cancer Cells HCT116. *Int. J. Mol. Sci.* **2018**, *19*, 2269. [CrossRef] [PubMed]
87. Vila, L.; García-Rodríguez, A.; Cortés, C.; Marcos, R.; Hernández, A. Assessing the effects of silver nanoparticles on monolayers of differentiated Caco-2 cells, as a model of intestinal barrier. *Food Chem. Toxicol.* **2018**, *116*, 1–10. [CrossRef] [PubMed]
88. Abdelkhaliq, A.; van der Zande, M.; Undas, A.K.; Peters, R.J.B.; Bouwmeester, H. Impact of in vitro digestion on gastrointestinal fate and uptake of silver nanoparticles with different surface modifications. *Nanotoxicology* **2019**, 1–16. [CrossRef] [PubMed]
89. Trickler, W.J.; Lantz, S.M.; Murdock, R.C.; Schrand, A.M.; Robinson, B.L.; Newport, G.D.; Schlager, J.J.; Oldenburg, S.J.; Paule, M.G.; Slikker, W., Jr.; et al. Silver nanoparticle induced blood-brain barrier inflammation and increased permeability in primary rat brain microvessel endothelial cells. *Toxicol. Sci.* **2010**, *118*, 160–170. [CrossRef] [PubMed]
90. Çiftçi, H.; Türk, M.; Tamer, U.; Karahan, S.; Menemen, Y. Silver nanoparticles: Cytotoxic, apoptotic, and necrotic effects on MCF-7 cells. *Turk. J. Biol.* **2013**, *37*, 573–581. [CrossRef]
91. Jiao, Z.H.; Li, M.; Feng, Y.X.; Shi, J.C.; Zhang, J.; Shao, B. Hormesis effects of silver nanoparticles at non-cytotoxic doses to human hepatoma cells. *PLoS ONE* **2014**, *9*, e102564. [CrossRef] [PubMed]
92. Khorrami, S.; Zarrabi, A.; Khaleghi, M.; Danaei, M.; Mozafari, M.R. Selective cytotoxicity of green synthesized silver nanoparticles against the MCF-7 tumor cell line and their enhanced antioxidant and antimicrobial properties. *Int. J. Nanomed.* **2018**, *13*, 8013–8024. [CrossRef]
93. Chernousova, S.; Epple, M. Silver as antibacterial agent: Ion, nanoparticle, and metal. *Angew. Chem. Int. Ed.* **2013**, *52*, 1636–1653. [CrossRef]
94. Yuan, Y.G.; Zhang, S.; Hwang, J.Y.; Kong, I.K. Silver Nanoparticles Potentiates Cytotoxicity and Apoptotic Potential of Camptothecin in Human Cervical Cancer Cells. *Oxid. Med. Cell. Longev.* **2018**, *2018*, 6121328. [CrossRef]
95. Sharma, H.S.; Ali, S.F.; Hussain, S.M.; Schlager, J.J.; Sharma, A. Influence of engineered nanoparticles from metals on the blood-brain barrier permeability, cerebral blood flow, brain edema and neurotoxicity. An experimental study in the rat and mice using biochemical and morphological approaches. *J. Nanosci. Nanotechnol.* **2009**, *9*, 5055–5072. [CrossRef]
96. Mwilu, S.K.; El Badawy, A.M.; Bradham, K.; Nelson, C.; Thomas, D.; Scheckel, K.G.; Tolaymat, T.; Ma, L.; Rogers, K.R. Changes in silver nanoparticles exposed to human synthetic stomach fluid: Effects of particle size and surface chemistry. *Sci. Total Environ.* **2013**, *447*, 90–98. [CrossRef] [PubMed]
97. Marchioni, M.; Jouneau, P.H.; Chevallet, M.; Michaud-Soret, I.; Deniaud, A. Silver nanoparticle fate in mammals: Bridging in vitro and in vivo studies. *Coord. Chem. Rev.* **2018**, *364*, 118–136. [CrossRef]

98. Cueva, C.; Gil-Sánchez, I.; Tamargo, A.; Miralles, B.; Crespo, J.; Bartolomé, B.; Moreno-Arribas, M.V. Gastrointestinal digestion of food-use silver nanoparticles in the dynamic SIMulator of the GastroIntestinal tract (simgi®). Impact on human gut microbiota. *Food Chem. Toxicol.* **2019**, *132*, 110657. [CrossRef] [PubMed]
99. Das, P.; McDonald, J.; Petrof, E.; Allen-Vercoe, E.; Walker, V. Nanosilver-mediated change in human intestinal microbiota. *J. Nanomed. Nanotechnol.* **2014**, *5*, 1.
100. Cattò, C.; Garuglieri, E.; Borruso, L.; Erba, D.; Casiraghi, M.C.; Cappitelli, F.; Villa, F.; Zecchin, S.; Zanchi, R. Impacts of dietary silver nanoparticles and probiotic administration on the microbiota of an in-vitro gut model. *Environ. Pollut.* **2019**, *245*, 754–763. [CrossRef] [PubMed]
101. Rahman, M.F.; Wang, J.; Patterson, T.A.; Saini, U.T.; Robinson, B.L.; Newport, G.D.; Murdock, R.C.; Schlager, J.J.; Hussain, S.M.; Ali, S.F. Expression of genes related to oxidative stress in the mouse brain after exposure to silver-25 nanoparticles. *Toxicol. Lett.* **2009**, *187*, 15–21. [CrossRef] [PubMed]
102. Jeong, G.N.; Jo, U.B.; Ryu, H.Y.; Kim, Y.S.; Song, K.S.; Yu, I.J. Histochemical study of intestinal mucins after administration of silver nanoparticles in Sprague-Dawley rats. *Arch. Toxicol.* **2010**, *84*, 63–69. [CrossRef]
103. Kim, Y.S.; Song, M.Y.; Park, J.D.; Song, K.S.; Ryu, H.R.; Chung, Y.H.; Chang, H.K.; Lee, J.H.; Oh, K.H.; Kelman, B.J.; et al. Subchronic oral toxicity of silver nanoparticles. *Part. Fibre Toxicol.* **2010**, *7*, 20. [CrossRef]
104. Shahare, B.; Yashpal, M. Toxic effects of repeated oral exposure of silver nanoparticles on small intestine mucosa of mice. *Toxicol. Mech. Method* **2013**, *23*, 161–167. [CrossRef]
105. Skalska, J.; Frontczak-Baniewicz, M.; Strużyńska, L. Synaptic degeneration in rat brain after prolonged oral exposure to silver nanoparticles. *NeuroToxicology* **2015**, *46*, 145–154. [CrossRef]
106. Xu, L.; Shao, A.; Zhao, Y.; Wang, Z.; Zhang, C.; Sun, Y.; Deng, J.; Chou, L.L. Neurotoxicity of Silver Nanoparticles in Rat Brain After Intragastric Exposure. *J. Nanosci. Nanotechnol.* **2015**, *15*, 4215–4223. [CrossRef]
107. Wilding, L.A.; Bassis, C.M.; Walacavage, K.; Hashway, S.; Leroueil, P.R.; Morishita, M.; Maynard, A.D.; Philbert, M.A.; Bergin, I.L. Repeated dose (28-day) administration of silver nanoparticles of varied size and coating does not significantly alter the indigenous murine gut microbiome. *Nanotoxicology* **2016**, *10*, 513–520. [CrossRef] [PubMed]
108. Williams, K.; Milner, J.; Boudreau, M.D.; Gokulan, K.; Cerniglia, C.E.; Khare, S. Effects of subchronic exposure of silver nanoparticles on intestinal microbiota and gut-associated immune responses in the ileum of Sprague-Dawley rats. *Nanotoxicology* **2015**, *9*, 279–289. [CrossRef] [PubMed]
109. Hendrickson, O.D.; Klochkov, S.G.; Novikova, O.V.; Bravova, I.M.; Shevtsova, E.F.; Safenkova, I.V.; Zherdev, A.V.; Bachurin, S.O.; Dzantiev, B.B. Toxicity of nanosilver in intragastric studies: Biodistribution and metabolic effects. *Toxicol. Lett.* **2016**, *241*, 184–192. [CrossRef] [PubMed]
110. Siczek, K.; Zatorski, H.; Chmielowiec-Korzeniowska, A.; Pulit-Prociak, J.; Smiech, M.; Kordek, R.; Tymczyna, L.; Banach, M.; Fichna, J. Synthesis and evaluation of anti-inflammatory properties of silver nanoparticle suspensions in experimental colitis in mice. *Chem. Biol. Drug Des.* **2017**, *89*, 538–547. [CrossRef] [PubMed]
111. van den Brule, S.; Ambroise, J.; Lecloux, H.; Levard, C.; Soulas, R.; de Temmerman, P.; Palmai-Pallag, M.; Marbaix, E.; Lison, D. Dietary silver nanoparticles can disturb the gut microbiota in mice. *Part. Fibre Toxicol.* **2016**, *13*, 38. [CrossRef]
112. Bacchetta, C.; Ale, A.; Simoniello, M.F.; Gervasio, S.; Davico, C.; Rossi, A.S.; Desimone, M.F.; Poletta, G.; López, G.; Monserrat, J.M.; et al. Genotoxicity and oxidative stress in fish after a short-term exposure to silver nanoparticles. *Ecol. Indic.* **2017**, *76*, 230–239. [CrossRef]
113. Merrifield, D.; Shaw, B.; Harper, G.; Saoud, I.; Davies, S.; Handy, R.; Henry, T. Ingestion of metal-nanoparticle contaminated food disrupts endogenous microbiota in zebrafish (*Danio rerio*). *Environ. Pollut.* **2012**, *174*, 157–163. [CrossRef]
114. Liu, H.; Wang, X.; Wu, Y.; Hou, J.; Zhang, S.; Zhou, N.; Wang, X. Toxicity responses of different organs of zebrafish (Danio rerio) to silver nanoparticles with different particle sizes and surface coatings. *Environ. Pollut.* **2019**, *246*, 414–422. [CrossRef]
115. Liu, X.; Dumitrescu, E.; Kumar, A.; Austin, D.; Goia, D.; Wallace, K.N.; Andreescu, S. Differential lethal and sublethal effects in embryonic zebrafish exposed to different sizes of silver nanoparticles. *Environ. Pollut.* **2019**, *248*, 627–634. [CrossRef]

116. Schultz, C.L.; Wamucho, A.; Tsyusko, O.V.; Unrine, J.M.; Crossley, A.; Svendsen, C.; Spurgeon, D.J. Multigenerational exposure to silver ions and silver nanoparticles reveals heightened sensitivity and epigenetic memory in *Caenorhabditis elegans*. *Proc. Biol. Sci.* **2016**, *283*, 20152911. [CrossRef] [PubMed]
117. Yang, Y.; Xu, G.; Xu, S.; Chen, S.; Xu, A.; Wu, L. Effect of ionic strength on bioaccumulation and toxicity of silver nanoparticles in *Caenorhabditis elegans*. *Ecotoxicol. Environ. Saf.* **2018**, *165*, 291–298. [CrossRef] [PubMed]
118. Moon, J.; Kwak, J.I.; An, Y.J. The effects of silver nanomaterial shape and size on toxicity to Caenorhabditis elegans in soil media. *Chemosphere* **2019**, *215*, 50–56. [CrossRef] [PubMed]
119. Wamucho, A.; Unrine, J.M.; Kieran, T.J.; Glenn, T.C.; Schultz, C.L.; Farman, M.; Svendsen, C.; Spurgeon, D.J.; Tsyusko, O.V. Genomic mutations after multigenerational exposure of *Caenorhabditis elegans* to pristine and sulfidized silver nanoparticles. *Environ. Pollut.* **2019**, *254*, 113078. [CrossRef]
120. Panacek, A.; Prucek, R.; Safarova, D.; Dittrich, M.; Richtrova, J.; Benickova, K.; Zboril, R.; Kvitek, L. Acute and chronic toxicity effects of silver nanoparticles (NPs) on Drosophila melanogaster. *Environ. Sci. Technol.* **2011**, *45*, 4974–4979. [CrossRef]
121. Araj, S.E.A.; Salem, N.M.; Ghabeish, I.H.; Awwad, A.M. Toxicity of Nanoparticles against Drosophila melanogaster (Diptera: Drosophilidae). *J. Nanomater.* **2015**, *2015*, 9. [CrossRef]
122. Raj, A.; Shah, P.; Agrawal, N. Dose-dependent effect of silver nanoparticles (AgNPs) on fertility and survival of Drosophila: An in-vivo study. *PLoS ONE* **2017**, *12*, e0178051. [CrossRef]
123. Alaraby, M.; Romero, S.; Hernández, A.; Marcos, R. Toxic and Genotoxic Effects of Silver Nanoparticles in Drosophila. *Environ. Mol. Mutagen.* **2019**, *60*, 277–285. [CrossRef]
124. Munger, M.A.; Radwanski, P.; Hadlock, G.C.; Stoddard, G.; Shaaban, A.; Falconer, J.; Grainger, D.W.; Deering-Rice, C.E. In vivo human time-exposure study of orally dosed commercial silver nanoparticles. *Nanomed. Nanotechnol. Biol. Med.* **2014**, *10*, 1–9. [CrossRef]

MDPI
St. Alban-Anlage 66
4052 Basel
Switzerland
Tel. +41 61 683 77 34
Fax +41 61 302 89 18
www.mdpi.com

Microorganisms Editorial Office
E-mail: microorganisms@mdpi.com
www.mdpi.com/journal/microorganisms

www.ingramcontent.com/pod-product-compliance
Lightning Source LLC
LaVergne TN
LVHW070842170726
843515LV00005B/557

* 9 7 8 3 0 3 9 3 6 5 5 1 7 *